Technology and the Lifeworld

The Indiana Series in the Philosophy of Technology

Don Ihde, general editor

Technology and the Lifeworld

From Garden to Earth

by Don Ihde

Indiana University Press
Bloomington and Indianapolis

The paper used in this publication meets the minimum requirements of American National Standard for Information Sciences—Permanence of Paper for Printed Library Materials, ANSI Z39.48-1984.

Manufactured in the United States of America

Library of Congress Cataloging-in-Publication Data

Ihde, Don, 1934-
    Technology and the lifeworld : from garden to earth / Don Ihde.
      p. cm. — (The Indiana series in the philosophy of technology)
      ISBN 0-253-32900-0 (alk. paper). — ISBN 0-253-20560-3 (pbk.: alk. paper)
      1. Technology—Philosophy.   I. Title. II. Series.
T14.I353   1990
601–dc20   89-45472
                                                                                                        CIP

    1 2 3 4 5   94 93 92 91 90

for Mark Hillel Ihde

Contents

Preface

Indiana University Press is proud to launch the Indiana Series in the Philosophy of Technology with the following trio: *John Dewey's Pragmatic Technology*, by Larry A. Hickman; *Technology and the Lifeworld: From Garden to Earth*, by Don Ihde; and *Heidegger's Confrontation with Modernity: Technology, Politics, Art* by Michael E. Zimmerman.

The Indiana series is the first North American series explicitly dedicated to the philosophy of technology. (There are other series relating philosophy and technology, particularly those which collect interdisciplinary articles, but none of these has been devoted to the development of a new subdiscipline within philosophy.) Broadly conceived, it will address a wide variety of issues relating to technology from distinctly philosophical perspectives. Philosophically, our approach will be pluralistic; this is apparent in our first round of books, which reflects both the traditions of American pragmatism and Euro-American trends.

Our *trio* is a timely one. We begin with radical reappraisals and new interpretations of the two early twentieth-century philosophers who made questions of technology central to their thought, John Dewey and Martin Heidegger. And we are also beginning with a systematic reformulation of a framework and set of questions regarding technology in its cultural setting in *Technology and the Lifeworld*. Later, we will be adding volumes of a more topical and thematic nature; subjects include "Engineering Birth," "Big Instrument Science," "Media and Rationality," "Technological Transformations of Perception," and many others.

Our goals are to include philosophically critical, historical, and interpretive studies as well as original and topical ones within a perspective which is balanced, reasoned, and rigorous regarding the newly emergent field of the philosophy of technology.

Technology and the Lifeworld is intended to set a tone for the series. It seeks to avoid the extremes of both utopian and dystopian interpretations of technology, which have too often captured the field. It draws from a very wide background of interdisciplinary and scientific studies. In this case, the twin foci are those of human-technology relations and of the cultural embeddedness of technologies. It is aimed at an educated, but wide, audience.

What has just been said is in an "editorial" voice. Because in this case the editor is also the author, I now take the occasion to change voices and describe more particularly the background to this volume.

The journey, both intellectual and geographical, behind this book has been a long one. In one respect, it goes all the way back to my Kansas boyhood in a farming community. My father, uncles, and their neighbors were all "technologists" in some sense. One of my uncles was an early flier (of the old bi-planes, which in higher-tech versions are still used as crop dusters), and a second cousin used to land in our pasture to give my brother and me rides in the old Aeronca. Everyone invented or dreamed of inventing something; and, as bricoleurs, my brother and I invented a machine to lift bales from the ground to the hayrack—a job which by hand we both disliked—and it worked. There was both a romance about machines and a sense of "it can be done."

This was also a time of transition from the community-owned threshing machine, which went from farm to farm to thresh the wheat, to the age of the mega-combine. The latter's relatives I have seen multiplied. Last fall in Italy's chianti country, I watched an equally large grape harvester pick the grapes for wine production; this demolished my belief that hand labor is still used in this process.

Philosophizing about technology, of course, came a long time after my farm boyhood. Its roots were perhaps taking shape while I was a graduate student at Boston University and working at MIT among the elite of the technology development faculty, but it was not until 1975, having been at Stony Brook for six years, that I published my first article, "A Phenomenology of Man-Machine Relations" (whose title was criticized by a reviewer for its "masculinist" bias—I thenceforth used "human-technology" relations as a better term). After that, my interest began to accelerate. In 1979 I published a collection of essays as the first volume on the philosophy of technology in the Boston Philosophy of Sciences Series, *Technics and Praxis* (Reidel). As late as ten years ago, philosophy of technology in North America could hardly be called a philosophical subfield. Indeed, in that same year the philosopher of science Mario Bunge wrote: "Technophilosophy is still immature and uncertain of its very object and does not exploit the entire scope of its own possibilities. That it is an underdeveloped branch of scholarship is suggested by the fact that so far, no major philosopher has made it his central concern or written an important monograph on it."[1]

In Europe, of course, there had been a much longer tradition. The first book using the title *Technikphilosophie* had been published in 1877 by the neo-Hegelian Ernst Kapp, who made his home both in Germany and in Texas. A German philosophy of technology society was several decades old by the early 1980s, when it was presided over by Friedrich Rapp (who wrote the second book on philosophy of technology in the Boston Series, *Analytic Philosophy of Technology*, 1981). Rapp and Paul Durbin, the "father" of the Society for Philosophy and Technology, which had been meeting for some time but which officially organized only in 1983, launched a select conference of German and American philosophers—twelve each—which was held in 1981 in Bad Homburg, Germany. That group, now expanded to a wider international circle, still meets on alternate years, usually alternating between the United States and Europe.

Perhaps it was natural that such societies should take their places within the high-technology, industrially developed universities of the Northern Hemisphere. It was in this Euro-American context that, until then, I had done my own work. By 1982, however, in part in response to a somewhat surprising reception of *Technics and Praxis* in developing areas, I began to receive invitations to Southern Hemisphere universities.

The first of these was the El Rosario Universidad (cooperating with three other universities) in Bogotá, Colombia, South America. There, in an intensive faculty-development seminar in February 1982 on the philosophy of technology, I received something of an awakening about the provincialism of my previous Euro-American context. Not only did I discover the striking difference between the South American context and assumptions being made about science and technology but I had to recreate my entire already planned program while I was in the process of responding to the new issues. The Euro-American worries about the relations between science and technology seemed like some kind of embroidery upon the more substantive issue of the cultural impact of a technologized science on the Latins. Since technology-science was seen as a single, unitary phenomenon, it spread widely through the cultural-political framework and was often perceived as having a negative influence on indigenous practices and values. It was then that I began to see technologies as *cultural instruments,* as well as the scientific instruments of my earlier work.

Later in the same year I visited a number of universities in South Africa. Refusing any government sponsorship but curious to see first-

---

1. Mario Bunge, "Five Buds of Techno-Philosophy," *Technology in Society* 1 (Spring 1979): 68.

hand this distressing part of the world, and receiving support through BMW's interest in philosophy of technology and the University of Zululand (a black university), I spent a month in South Africa. The attitude there was very different from the fairly negative one in Bogotá. The theme of the main conference and lecture series was "Technology and Utopia," and in many sectors technological science was perceived as the means to leapfrog into the twenty-first century.

More recent still was a trip to New Delhi, India. Here, in a country in which there is still lively debate about a "Gandhian" or a "Nehruic" direction, there is at least agreement about retaining national autonomy of development in the context of a quasi-socialistic approach, with an official policy of rapid and positive science-technology development. The National Institute for Science and Technology Development is an impressive research institute, staffed by some of the best young minds in India but faced with one of the most enormous tasks imaginable in the face of population extrapolation and continued poverty.

These experiences and others helped convince me that a book attempting to reframe the question of technology must take account of the multicultural—that is, the more international—setting within which technological culture is taking and will take place. That is the reason for the particular emphasis here, particularly in the second and third parts of the "program" I have devised.

My choice of focus inevitably leaves other dimensions of the technological lifeworld underdeveloped. There are gaps, the largest and most important of which is the social-political dimension. Although this dimension is not entirely absent, it remains in the background. Similarly, the sociology and politics of technological science itself are underplayed. I have not even tried to cover all the frontiers of emergent technologies. The most conspicuous absence is a more extended discussion of biotechnology, clearly of crucial importance to the future of technology. I do not claim comprehensiveness but rather a thematic focus that relates science and technology to its human experiential and cultural dimensions.

I have also been quite transparent about the role of my own experience in the book. That has always been a feature of my style, and I recognized that some readers like and others dislike such autobiographical vignettes. Although writers are guided by and respond to their own experiences, I have sought both to do the groundwork that would make my claims serious ones and to do the homework in the usual academic fashion, through careful and avid reading, to a point beyond anything simply autobiographical. This is the case with the multiple cultures used as examples here as well. From early graduate programs days at Stony Brook (where our curriculum requires a year of team-taught interdisciplinary work, in which I shared philosophy-anthropology interfaces) to the trip to the Australian National University in 1985 to study further both the aboriginal and South Pacific groups I refer to, I have sought to examine and verify what may be claimed for these cultures.

There are two more experiences which are relevant here. The first is the cause of the delay in getting the book done on its original schedule—the acceptance of a deanship for the Humanities and Arts at Stony Brook in September 1985. Although that job precluded completion of the manuscript until a research leave in fall 1988, it provided me with much firsthand experience in the "sociology and politics" of science and technology faculties. Stony Brook, typical of large public research universities, is not only dominated by the (natural) sciences but is administered at the highest levels by a team of (natural) scientists-turned-administrators. From this situation I have learned many lessons, only a few of which are reflected here.

The other experience puts "you are what you eat" in the center of so many of my tales. In recruitments for the Philosophy Department, we used to joke that after determining the individual's philosophical promise, the "real" questions would begin, among them "What cuisine do you practice?" in the hope that a new one could be added to the social repertoire.

The feast as a setting for philosophizing is at least as old as the *Symposium,* although I have been a bit more carried away with what is served in my effort to make metaphorical points. This runaway metaphor may also be due, in part, to the final writing in the hills of Tuscany, Italy, in the land of one of the world's great cuisines. That choice of location— made more out of typical academic need to negate the daily pressures of one's home university than for other reasons—was not the only rationale.

Tuscany also had fortuitous historical and philosophical implications, for it is the area of the birth of Renaissance science, modern science. Vinci, Leonardo's village, was just around the corner from my Tuscan farmhouse, and Pisa, Galileo's setting, was but an hour away. This circumstance gave an unexpected concreteness, unplanned but now appreciated, to the writing process.

After this rather long gestation, *Technology and the Lifeworld* is now finished. The early stages of this book were supported by a sabbatical spent in Paris in 1984, followed by a summer as a Visiting Fellow at the Australian National University, Canberra, 1985, through the auspices of Richard Campbell and the Faculty of Arts. The research semester in Spazzavento, Italy, in fall 1988, was again supported by Stony Brook.

A book takes shape within an intellectual community. Here, too, the project began at Stony Brook but reached out to the corners of the globe. I always have second and third thoughts about whom to mention and worry about who should have been mentioned and in what contexts. In the present circumstances, I will simply name some of the individuals who quite directly contributed in the critical discussions and in print and to whom I owe special gratitude: At Stony Brook there is Patrick Heelan, along with Marshall Spector, Ruth Cowan, John Truxal, and, more recently, Ann Kaplan. Then, too, I must mention the two successful completers of doctoral degrees in philosophy of technology during the birth pangs of the book, Paul Thompson and Larry Kilbourne.

Very careful attention and criticism was made by Robert Ackermann, Albert Borgman, John Compton, and Frederick Ferre, with extended discussions with Robert Cohen, Paul Durbin, Lester Embree, Larry Hickman, Joseph Kockelmans, Lenore Langsdorf, John McDermott, Joseph Pitt, Langdon Winner, and Marx Wartofsky and, recently, Sandra Harding, Susan Squier, and Anthony Weston.

The list outside North America is equally large, and I should cite Friedrich Rapp, Richard Sylvan, C. S. DeBeer, Maurita Harney, and R. Sinari among these.

Of course, none of this would have happened at all, particularly among the multiple pressures and pulls of travel, deaning, and the rest, were it not for my supersecretary, Jean Kelley. She was able to overcome the premodern Italian mail system and get the entire manuscript reduced to disks within the semester of research assignment and the intersession following. Then, at nights and over weekends, she did the first copyediting on my various punctuation and insert foibles. I only hope she did not suffer from the new computer "disease" I discovered in Australia called "repetition strain injury," which occurs with too-long hours in front of a VDT. Other hours were put in by Curt Naser of the Humanities Institute checking footnotes and technical details. And Mary Jane Gormley, Indiana's copyeditor, not only corrected and trimmed excess prose but did so with the best sense of humor and added insights in my long experience with persons doing this task.

At home, my wife Linda Einhorn-Ihde, with her background in English, willingly lent her skills in grammar and style. And my young son, Mark Hillel Ihde, to whom the book is dedicated, provided both a new sense of enthusiasm and a reason to see, through him, the hopes for the directions our technologies might take in the future.

Introduction: Entry Level

By this late part of the twentieth century, those of us who live in the industrially developed parts of the Northern Hemisphere live and move and have our being in the midst of our technologies. We might even say that our existence is *technologically textured*, not only with respect to the large dramatic and critical issues which arise in a high technological civilization—such as the threat of nuclear war or the worry over global pollution, with its possibly irreversible effects—but also with respect to the rhythms and spaces of daily life.

If we begin this inquiry with a broad, but also concrete and experiential, notion of technologies as those artifacts of material culture that we use in various ways within our environment, we could begin a daily calendar of our human-technology interactions that would prove numerically staggering. Imagine a most mundane, diary-like approach, which might begin with our first waking moments:

- It is likely that we are called into waking consciousness by a technology, be it the ringing of an alarm, the beeping of a quartz clock, or the sounds of a clock-radio.
- A first survey made as we reluctantly leave the warmth of the bed would reveal that our sleep had been secured by our sleep technologies (whether of the higher sort—an electric blanket—the medium sort of acrylic blankets, or the low sort in which natural fibers nevertheless have been submitted to a variety of transformations implicating technology).
- Waking becomes even more enmeshed with complex technological texture the moment we move to the bathroom. A vast plumbing system is engaged, the history of which stretches back to the Etruscans and Romans but which today entails vast water systems and metallurgical and plasticized materials.
- Then we breakfast in a kitchen probably dominantly electrified and even more in contrast with kitchens of an even longer history. The automatic machines (toaster, coffeemaker—probably preset for our arising—microwave oven, dishwasher, etc.) allow us quickly to breeze through this first repast of the day.
- After awakening and eating, it is likely that we leave home by a transportation system with intricate and extended involvements in systems;

our new automobiles even report to us on their conditions through the computerized and electronically displayed instruments which bespeak their kind of human-technology communications.
• All of this mundane activity will probably also be accompanied by other dimensions of human-technology interaction. For example, while eating, the television might have been on to let in the world; while driving to work, the radio entertains or informs; if the driver is more a workaholic, the dictating machinery might already be engaged.

This first hour of waking existence is already sufficient to demonstrate how thoroughly intertwined our activities are with technologies. Let us take two less ordinary examples to take account of the same textural involvement:

Sometimes recreation takes the form of an escape from the intense saturation of urban and material culture. For instance, one might wish to retreat to one of our many "nature museums," enclosed but protected wilderness areas, parks, and the like. One challenging form of such recreation might well be mountain climbing. The climber will employ pitons and carabiners of metal alloys, climbing ropes of braided dacron, specially manufactured climbing shoes, and for an overnight climb, probably a lightweight, artificial-fiber tent and miniaturized cooking equipment and dehydrated foods.

In a more intimate context, the same observations could be made about twentieth-century sexual practice. What contemporary person engages in sexual relations without *some* decisional relation to birth-control technologies? The simple awareness that there are such techniques involving technologies has changed the context of sexual practice. The "decision" *not* to use such devices is taken in the light of their known existence; and in fact in most cases it is precisely the failure to use rather than the now-assumed use that draws social disapproval. At a demographic level, the small size of families in most industrial societies demonstrates what the normative practice is.

In our era, in contrast to most traditional societies, the decision *to* conceive is the exception to the norm of *not* conceiving, and the responsibility for conception is clearly related to our knowledge and use of birth-control technologies. This may be well illustrated by what we take to be a major social problem in high-technology societies, that is, adolescent pregnancy. First, adolescent pregnancy is considered a problem in most such countries in contrast to what is in fact the norm in many other cultures of a traditional sort. Second, the problem arises not so much because of the acceptance or non-acceptance of adolescent sexual activity as of the "failure" to engage in such activity "responsibly." To the shock of many first-time teachers and social workers participating in programs dealing with adolescent pregnant girls, what is usually discovered is that there was knowledge of birth control but a kind of "decision"—often through behavioral passivity—has been taken *not* to use contraceptives. Whatever the reason, the societal expectation is that adolescents are responsible only if they prevent possible pregnancies, which is to say that responsibility entails using available technologies.

All of this is familiar, including the range and extent to which our daily activities are involved with technologies. But simply because of its familiarity, we may overlook both the need for and the results to be obtained by a critical reflection upon our lives within this technologically textured ecosystem—perhaps better termed a *technosystem*. Indeed, the very familiarity might allow us to be taken aback were anyone to characterize this life as in any way peculiar.

Yet a serious reflection can only begin by gaining precisely enough distance from our mundane involvements that some sense of their uniqueness—even peculiarity—can be grasped. There is a large range of important, existential questions lurking within the very taken-for-granted realm of our daily activities. Nor do we lack a common awareness of some of these problems, which take the shape of popular beliefs on one level but which have counterparts even within the most sophisticated reaches of expertise. The following questions and the accompanying debates about them are indicative:

How like or unlike is life within our technosystem from previous or other forms of life that humans have taken up? Popular and expert consciousness offers a mixed response to this question. On one side, the perhaps dominant belief is that contemporary technologies are somehow essentially and dramatically different from all past technologies. This means that there must be some fairly dramatic disjunction between us and the past and between those in highly industrialized, high-technology societies and those not yet within such a technosphere.

But is this the case? And, if so, in what respects? Robert Oppenheimer, the Manhattan Project physicist, is said to have observed that there are more physicists alive today than in all previous centuries (to which the humorist Art Buchwald has given the properly critical and philosophical response, "At that rate, there will soon be two physicists for every person alive!"). This quantum extrapolation is often cited with respect to various "explosions" (information, population, knowledge, physicists, and the like). It hints that such a jump in quantity implies a corresponding leap in quality, but its aim remains a reenforcement of the belief in a disjunction of scientific and technological culture with previous cultures.

Associated with this belief is another—in this case, often held at the expert level. It is the belief that the crucial factor in creating the disjunction is science, particularly in its theoretical form. Within the narrower communities of philosophy of science, among many scientists, and even among some historians, the dominant belief is that contemporary technology is the result of science, either chronologically or, more often, because of the dependence of application from theory.

The relationship between science and technology has often been a central question in North American and European circles concerned with reflections upon technology. That technology is *applied science* remains the standard or dominant view. And such technologies are to be strongly contrasted with the handwork or craft technologies of the past.

This dominant view has increasingly come under attack. At the least, it is now open to question. What is the relationship between science and technology? And what are the implications of those relations both for the special cases of scientific knowledge and, more broadly, for the full range of our knowledge activities?

A second set of questions, equally important, revolves around whether technologies are neutral. Are technologies *mere* things that, like inert matter, do nothing in themselves? Or do technologies affect the very ways we act, perceive, and understand? At the popular level, the belief that technologies are mere artifacts-in-themselves is sometimes proclaimed in bumper stickers: "Guns don't kill people, people kill people." But the belief may also take shape in expert consciousness that technologies do not do anything; it is only how they are used.

For example, debates revolving around technological development may take the shape of an argument between those who take a social-determinist position and those who take a technological-determinist position. Social scientists frequently take some form of the first position, arguing in effect that what really counts in technological development is some set of decisions by a power elite. Technologies, in such a view, are resultant from such decisions (in this case, the power elite substitutes for the theory elite of the previous science/technology argument) and technology is the result. In the perspective of social determinists, the particular technologies usually remain background factors against which the human social and political conflicts take shape.

A good example of this position may be illustrated by the development of the McCormick Reaper. The reaper was a technology which replaced the hand scythes of a number of farm workers. Its sickle bar cut a wide swath of grain and bound the grain into bundles. The cutting bar was mechanically driven by a complex set of cast gears that, in the beginning, were made by skilled craftsmen who produced the molds for casting, did the casting, and finished off the rough products.

Management, however, soon found the demands of the artisans too high in labor costs, so a new technique was developed that broke down the process into individual steps such that any unskilled person could learn one such step and practice it. This was typical nineteenth-century de-skilled practice designed largely to disenfranchise nascent unions.

The result was quite counter to any technological imperative toward efficiency and cost. The production of gears by such de-skilled factory processes was both slower and more expensive than the previous artisannal practice, thus making reapers more expensive. This management-production practice did achieve one aim: the elimination of the union. Then the McCormick factory returned to its former production practice with newly trained artisans.

Insofar as the analysis is socially deterministic, one could note simply that the process and product of the technology takes a secondary role in the background of management/labor controversy and that whatever technologies are ultimately involved make little difference to the case itself. Such a perspective simply does not deal with the technologies *per se*. The functional effect is to have interpreted technology as neutral.

The technological determinist, on the other hand, finds in the development of a technology itself a whole nest of possibilities that determine future directions for the socius. A good example comes from James Burke's reworking of Lynn White's earlier work on medieval technology.[1] Both Burke and White saw in the development of the stirrup a crucial technological juncture that affected the later social development of Medieval Europe. Once the stirrup was invented or discovered (it may have been borrowed, as so much medieval technology was, from Eastern sources), a new style of horseback riding became possible. A warrior, instead of using his lance to plunge down in a stabbing motion at a foot soldier, could now charge at speed, using the power of the horse to impale the enemy. A whole gestalt of changes could and did occur: Lances changed shape—they had to have a cross piece or a tie to prevent going so deeply into the victim as to be hard to withdraw. The stopper or cross piece and the colorful band (which later became individualized to identify the owner) were invented. The saddle became longer and was outfitted with a holding device for the lance. Then came armor, the breeding of larger horses, and the emergence of an elite cavalry (knights) and higher costs until the heavily armed, armored, and expensive "tank" of the Middle Ages resulted. Later, another technology, the longbow, which could serve as longer-range "artillery" and which was even capable of piercing armor, decreed the end of the cumbersome knight-horse combination.[2]

Here the technology clearly occupies a foreground position, and

---

1. Cf. James Burke, Connections (Boston: Little, Brown and Co., 1978), and Lynn White, Jr., Medieval Technology and Social Change (Oxford: Oxford University Press, 1962). The former popularized the latter and followed a historical set of often accidental connections through the history of Western technology in a documentary series on public television. I shall sometimes use illustrations from documentaries and widely available mid-range sources as a device for drawing upon familiar background. High-quality documentaries are, in fact, an essential part of the current mode of public science education and serve to dramatize much such investigation.

2. Burke, Connections, pp. 52–54.

equally clearly, the technology is not central with respect to either action or effect; but there is the suggestion that, once invented, technologies simply follow a line of development almost contextless, as it were.

At an even greater extreme of the neutrality/non-neutrality debate, there are those who hold not only that technologies are not neutral but that once created and put in place, technology (often with a capital: Technology) takes on a life of its own and becomes *autonomous*. The 1960s and 1970s brought to popularity a series of largely dystopian books that argued that Technology had outstripped human controls and, like the Frankenstein myth, was runaway. Two of the most widely read such books were Herbert Marcuse's *One Dimensional Man* and Jacques Ellul's *The Technological Society*. Both in effect argued that technology was equatable with certain calculative ("rational") *techniques* which in turn became not only dominant but total, thus making Technology as culture absolute. Ellul, in addition and in company with many other writers, contrasted this cultural milieu with the previous milieu, which he characterized as that of nature. "Technique has become the new and specific milieu in which man is required to exist, one which has supplanted the old milieu, viz., that of nature."[3]

With this interpretation of technology, another popular belief is raised: that technology by being produced is *artificial* and the artificial is to be contrasted with the *natural*. Such a belief may take very popular shape, for example, in the practices of avoiding foods with additives all the way to the pure use of only "natural" fibers in clothing. Or it may take sophisticated shape in philosophers' arguments such as those often made by Hans Jonas concerning the way technologies affect even the essence of humanity. Jonas argues that the penchant to seek power over nature makes for a "narrowness which is ready to sacrifice the rest of nature to [man's] purported needs, [and this] can only result in the dehumanization of man, the atrophy of his essence, even in the lucky case of biological survival."[4] The implication is that technology, by being inserted between humans and nature, could even change the essence of humanity (or what was once "human nature").

Dystopian interpretations of technology were popular in the last couple of decades, but utopian ones were so in the previous century. Some went so far as to extol the aesthetic properties of smokestack industries in that the smoke helped to create more beautiful sunsets! Then utopians could have been termed globalist in hope. Science-tech-

---

3. Jacques Ellul, The Technological order: Proceedings of the Encyclopedia Britannica Conference, ed. Carl E. Stores (Wayne: Wayne State University Press, 1963), p. 10.

4. Hans Jonas, "Responsibility Today: The Ethics of an Endangered Future," *Social Research* 43 (Spring 1976): 84.

nology, rightly applied and developed, it was believed, would eventually solve most, if not all, human social and personal problems. Certainly the major ones like poverty, crime, disease, pests, and the like would once and for all be eliminated. Today one seldom finds such globalistic utopianism. But what might be called specific or single system utopianism still abounds in various beliefs in the *technological fix*.

The top candidates for technological fix utopianism in recent times have been the "fundamentalists" of artificial intelligence (AI). As Hubert Dreyfus has so bitingly pointed out in his *What Computers Can't Do*, these apologists have been fast to feed the journalist's mills. In 1970, *Life* magazine claimed, on the basis of what AI apologists claimed, that there already were computers that "saw," "understood," "learned," and "thought." And "distinguished computer scientists are quoted as predicting that in three to fifteen years 'we will have a machine with the general intelligence of an average human being . . . and in a few months [thereafter], it will be at genius level.' "[5] One good result of the home computer revolution is that computers are now much more familiar to a wider set of users, with the predictable result that ordinary use has made this "expert's" mystique much less telling; today's AI community, while still inhabited by enough AI fundamentalists to keep the mythology alive, has largely become more modest about what metaphorical modeling processes seem promising.

More seriously, however, single system utopianism now has taken hold in much bionic–health science activity. From clearly helpful prosthetic devices and experimentation thereon (mini-radars for the blind and crude "reading" machines, computer-enhanced artificial limbs for the paralyzed, etc.), the high-technology medical experimenters have turned to internal bionics (artificial kidney, artificial heart—although the artificial brain still seems far off!). Lying embedded in such experimentation is a clear indication of beliefs about technological fixes vis-a-vis human-technology relations. What is the range of such relations? And what experiential structures do they display?

More serious still is the current experimentation on a technological fix which would presumably shield countries against nuclear attack: Star Wars. Here the economic and political implications are genuinely staggering and, if enacted, commit future generations to a questionable political-technological policy. Again, the assumptions within this political movement regarding the megatechnological fix are basically utopian.

A third set of existential questions involves the future. What does high technological development portend for our species' future? The popular and expert beliefs display a range of evaluations from dystopian to utopian. If technologies in the ensemble could be grasped—

---

5. Quoted in Hubert Dreyfus, What Computers Can't Do (New York: Harper and Row, Publishers, 1972), p. xxviii.

and even the question of the possibility of such an evaluation is difficult—would the ensemble indicate a Frankenstein phenomenon such
as dystopians hold? Or will the future become stable, and will the current set of problems that have emerged because of past scientific and
technological progress be solved? As we have emerged from earlier
eras, the catastrophes of nature, while still a part of human life, have
often been replaced by threats much more related to human activity
itself, that is, human technological activity. Humans-with-technologies
have clearly become a global force. Oceanic and atmospheric effects
caused by industrial practices are clearly indicators of this fact. So often the question about futures takes the shape of the *control* of technology. The very fact that the dominantly nineteenth-century question
of the control of nature now is placed side by side with the question
of the control of technology is itself suggestive of this shift in sensibility.

Can technology be controlled? Or has it become autonomous and
out of control? If controllable, how is it to be controlled? And under
what forms of authority? Questions about the future contain our fears
and hopes. The other form this set of questions takes is whether there
is some single overwhelming trajectory to technology as a whole. Will
there be a single, universal technological civilization? If so, will it be
democratic or totalitarian? Centralistic or decentralistic? Will it contain
variety or be monolithic? The literary play with such questions is evidenced in much of our cinematographic science fiction as well as in a
whole genre of books on the presumed effects of technology.

One does not have to turn to speculative literature to take note of
divergent present directions. In the debates surrounding nuclear energy production, one fear has been that high technologies contain a
determination toward centralization and hierarchical authority. Both
security (from nuclear products usable by terrorists to the production
of by-products radioactive for generations) and safety needs seem to
call for highly controlled and secretive authority structures. At a more
common level, the rapid introduction of computers into the home and
marketplace has made us aware both of the large potential for communication and exchange (direct lines to our banks, working at home
while in contact with the office, etc.) and threats of the invasion of privacy and massive breakdowns and errors. Many of us have gotten
caught in stubborn computer billing problems. All of us receive new
masses of junk mail because of the dissemination of computer lists by
advertisers.

On the world scale, traditional cultures are now rapidly disappearing. Does the march of contemporary technology determine the end
to this variety of human forms of life? Anthropologists are today scurry-
ing about to previously isolated cultures to record what many believe
to be their dying moments before absorption into the higher-technol-
ogy cultures of Western origin.

Even more prevalent are larger questions of whether our generation or the next one will undergo or survive a nuclear holocaust, or whether the longer-term effects of industrial technologies will produce some irreversible and possibly deleterious climatic effect, making life as we know it different, difficult, or impossible.

These questions and the familiar responses of both popular and expert consciousness would seem to call for serious philosophical critique. Yet philosophers, particularly those in North America, have been few and late in making technology a central theme of inquiry. Indeed, what is now belatedly becoming known as philosophy of technology has only recently arrived upon the scene. There are reasons for this late entry, but for the moment, the sheer magnitude both in scope and in depth of the questions should be awesome enough to give even the most ambitious philosopher pause. Yet it is precisely the aim of this book to enter the arena which can now be called the philosophy of technology.

It should be understood from the outset that the task of a philosophy, no matter how far-reaching or profound, is also limited. The philosopher cannot provide formulaic answers to the questions posed, nor are there in any likelihood such simple answers. There are two things that a philosophy can do: It can provide a perspective from which to view the terrain—in this case, the phenomenon of technology, or better, the phenomenon of human-technology relations. Second, a philosophy can provide a framework or "paradigm" for understanding. Both of these general philosophical aims will be undertaken here.

The first of these tasks is to establish a perspective from which to view the vast and complex terrain of technology and its human context. I suggested at the outset that part of that attainment of perspective calls for the right amount of distance, such that the uniqueness and even peculiarity of our technological culture could be seen. For insofar as critical thinking is like seeing, both that which is too close (the tip of our nose) and that which is too far (beyond the horizon) are simply not clearly discernible.

The usual or dominant perspective philosophers frequently choose might be called Lucretian. In his classic *De Rerum Natura*, Lucretius describes the philosophical perspective as one from a high (and, in some translations, "ivory") tower, from which the woes and movements of humans appear as distant and trivial, like the domain of ants to the standing human. In this distancing—long a metaphor for objectivity—is betrayed the preference for a position both fixed and distant. In a contemporary analogue, we might choose a satellite view. Just as today we have begun to take "whole earth measurements" via satellite, so the philosopher might prefer this high distance.

Yet in this case of the examination of technological culture, there is something radically wrong with such a choice. No more than we can take a pretended Martian view of ourselves can we escape a certain hidden closeness by the satellite metaphor. To the contrary, inhabiting the satellite would better illustrate precisely the enmeshment, the enclosure we have in our technosphere. The space vehicle is but one extreme example of a technological cocoon that must provide us with life support, a constructed environment, and the enclosed layer by and through which we would be taking our sightings. Far from giving us distance upon the phenomenon, we end by simply taking it for granted.

For that reason, and others to be indicated later, I prefer to take a more Kierkegaardian metaphor as the starting point for establishing a perspective. In *Fear and Trembling*, in the context of describing the inescapability of decision, Kierkegaard pictures us as captains of ships at sea, under way. The person at the wheel is already in motion; and to come about or not to come about is equally to make a decision. I shall modify this image by noting that the position of this perspective is basically *navigational*. The navigator, in the very midst of the sea where both boat and sea are in motion, must take bearings, find a direction, and locate both himself and his destination. This perspective occurs in a dynamic and fluid situation and is necessarily relativistic, yet just such a situation is normal for the navigator.

The navigational perspective is quite self-consciously aware of being in the midst of what is occurring, but the navigational problem is to locate reference positions through some means of variations. Here there is at least one good lesson that can be learned from space travel as well. Astronauts living in the various cocoons we have devised find themselves weightless. Their earthbound reference point, gravity relating to our normally upright posture, was missing. They had to *choose* a reference point from which to move. Frequently they would do this arbitrarily: "That knob there will mean 'up,' and that lever there, 'down.'" In short, they had to relearn a kind of bodily motility in a new frame of reference, and thus they invented a new kind of bodily navigation. Similarly, in the various styles of navigation, some means of referencing and some set of variations have to be established to attain the relativistic locations crucial for finding our way about.

Although I shall soon develop these suggestions in a more technical and precise way, the launching of the inquiry can begin quite informally, again with a device long familiar to philosophers. A contemporary myth or tale will serve, not unlike the story of the Cave invented by Plato to outline his epistemological theory, to point the way to what is to follow. In this case, however, I shall draw quite purposely on perhaps the most familiar myth in our traditions, the myth which divulges our appetites for innocence and our worries over leaving paradise—the story of the Garden of Eden.

1. From Garden to Earth

At entry level one unasked but tacit question was: Could humans live *without* technologies? Clearly, in any empirical or historical sense, they in fact do not. There are no known peoples, now or in historic or even prehistoric times, who have not possessed technologies in some minimal sense, yet we might still want to say that they could live so as an imaginative limit-possibility. An imaginative leap can be made to portray just such a form of life, a leap that will serve a recurring heuristic purpose.

We should from the beginning, however, be aware of the imaginative and even quasi-mythic quality of such an exercise. Since we envision it from the familiar and engaged position which we actually occupy within our more saturated technological form of life, we may not even be aware of just how deeply we are enmeshed, even at the perceptual level, in this form. Were we to return to our mountain climber with the pitons and carabiners, we might note that the very perception he or she might have of a mountain has already had a long and distinctive history that presupposes aspects of technological culture.

In the Middle Ages, for example, mountains were usually perceived as threatening, ominous, dangerous—and there is even some evidence in art history that they were rarely thought of as beautiful.

> Until the eighteenth century, for example, the Derbyshire peak region in England was considered wild and unfit for human eyes. In 1681, the poet Charles Cotton described it as a "country so deformed" that it might be regarded as "Nature's pudenda." Travelers in those days were advised to keep their coach blinds drawn while traversing the region so as not to be shocked by its ugliness and wildness.[1]

Yet by the nineteenth century mountains had become mystically beautiful (in the large landscape paintings of Europe), and one fad had viewers of mountains going about looking at them using wire frames to aesthetically enframe them. This reversal is surmised by Rene Dubos

---

1. René Dubos, *The Wooing of Earth* (New York: Charles Scribner's Sons, 1980), p. 14.

to have been partly a reaction to the rise of machine technology itself.
"The European pro-wilderness movement gained momentum in the
nineteenth century from the reaction against the brutalities of the Industrial Revolution."[2] Similarly, in Japan, the Japan Alps were considered ominous and the home of demons before European adventurers
began to explore and ski them; they have now become the recreational ground for Japan's large ski industry. The mountain climber, not
unaware of danger, nevertheless sublimates danger to the perception
of challenge, since few peaks are beyond the limit of the climbing now
made possible by climbing technologies. In an even more technologically textured observation, we might note that mountains today are
regularly "conquered" by roads, trails, and especially by overflight via
aircraft. The very experience of mountains has been transformed.

It is just for this reason that the New Garden must be imaginative,
for we cannot tell deeply what such a Garden would mean for our experience. It remains a story, an imaginative variation. Imagine a *New
Eden*, a new tale of beginnings, in which a New Adam and a New Eve,
like the old, appear first, naked and placed in the non-technological
Garden. Today the telling of such a tale should take on trappings familiar to our own context, perhaps the context of the well-informed television documentary viewer, itself the result of our scientifically
permeated society. The quest for the New Eden today will be guided
by acquaintance with the disciplines of anthropology, ecology, animal
behavior, geography, and the like.

Were we to find such anthropoids, we would know in advance
what humanity-markers to look for. The New Adam and New Eve, for
example, would clearly be language users—but they would obviously
be restricted to oral speech and its associated behaviors (gesture, body
language, etc.), although they could well have also developed their expressivity into song, poetry, dance in both aesthetic and religious
senses. But none of these activities could involve artifacts in an actional technics. (I shall often use the term "technics" to suggest human
action employing artifacts to attain some result within the environment.) Thus writing, musical instruments, masks, etc., would not be
employed.

Similarly, humans in the Garden could have complex mating and
sexual patterns which would preserve stability of families and kinship
lines. They could have traditions about marriages, multistable variants
such as those anthropologists are familiar with (patriarchal, matriarchal,
monogamous, polygamous, etc.), although these would be conveyed
orally and through the dramatic histories already suggested. But there
would be no records, calendars, or markers for the movement of heavenly bodies and the like.

This naked and face-to-face existence, however, also necessarily

2. Ibid., p. 15.

places ecological restrictions upon just where such a Garden could be. Because material culture (technology in the very broadest sense) could not exist, our primal pair must be found in a tropical region where temperature extremes would not cause hypo- or hyperthermia (no fire, no clothing, no constructed dwellings). The food supply must be constant and easily available. In a tropical environment, there would be fruit, edible plants, easily hand-caught frogs, fish, grubs, and the like. But our non-technological pair would not have storage technologies (baskets, pots) or hunting technologies (nets, spears, hooks) or even cooking technologies (fire and food preparation). Nor must we forget that our tropical paradise should be located somewhere where there would be an absence of large predators such as might endanger humans (no weapons either for defense or offense).

What this initial imaginative exercise reveals is that it might be possible for humans to live non-technologically as a kind of abstract possibility—but only on the condition that the environment be that of a garden, isolated, protected, and stable. The price for such a non-technological existence is to be enclosed. Here would be the "milieu of nature" in purer form. But there is no such empirical-historical human form of life because, long before our remembering, humans moved from all gardens to inherit the Earth.

The purpose of this exercise, however, is neither to introduce a nostalgia for the Garden nor to romanticize a presumed "natural" existence. Rather, the purpose here is to begin to take a measure upon the range of variations within which humans shape their forms of life. By beginning with this abstract variation, there is at least a suggestion that can be followed with respect to actual approximations to the face-to-face situation described. In tropical regions, there are still certain technologically *minimalist* cultures.

The remarkable Tasaday, a Stone Age culture first reported in 1972 in the backlands of Mindanao, are a good example of this. The Tasaday live in caves amidst a rich environment that provides them with a stable, if basic, existence. Yet they are hardly non-technological, in spite of the minimalism that characterizes their form of life. Take note of some basic aspects of this existence:
• Their diet consists of tadpoles, small crabs, frogs, abundant fruits, and certain vegetable matter such as palm hearts and other edible stalks. But food preparation is itself a technique which involves important, if basic, technologies. They make fire (friction method with two sticks, a technique employing artifacts), cook certain foods (wrapping tadpoles in large leaves and baking), and undertake other culinary preparations making for a result (cooked) materially different from its natural state (raw).
• Clothing is genuinely minimal, but adults wear vegetable-matter loincloths.
• Temperature control is purposely employed (fires in the caves both to cook and to ward off cold night air).

• While there is no sophisticated weaponry, the Tasaday do have a rat-tan-bound stone axe, which is used not to kill prey but to crack nuts and break hard-to-open foodstuffs.
• On a more macro scale, they have traps for deer, sluices for water, and methods for securing starch from palm pith, all involving fairly complicated techniques and technologies.

This people, first encountered in 1972 under the auspices of a *National Geographic* exploration, seem such an anachronism that the suspicion has been expressed that they were a "plant" of the Philippine government (subsequently denied by *National Geographic*). Yet the point is one which can hold here: Whether the Tasaday had ever left their "Garden," they had long since left the mode of non-technological existence and inherited their local Earth with at least minimal technologies. They belong to the same movements which took humans into the extreme corners of the earth, reaching even to the Arctic regions. The Inuit (Eskimo) developed a basically Stone Age technology to a high skill:
• Their diet is radically different from that of the Tasaday. Much more meat (whale, seal, fish, bear—often fatty for protection against the cold) is eaten, along with seasonal vegetable foods.
• To attain this diet, a hunting culture developed with very sophisticated technologies derived from animal material. Sinew, bone, fur serve a multiplicity of purposes ranging from weapons (harpoon, bow and arrows, spears) to vessels (the sophisticated kayaks, umiaks, and other boats, some of which arrived with live passengers in medieval Ireland after being swept across the Atlantic in storms!).
• Housing to be used in the migratory patterns was a skin and frame shelter for warm weather and the snow igloo for winter (complete with whale oil lamps, an ice window for light, and an ambient temperature warm enough to go unclothed).

There are similar ingenious adaptations, with simple and minimal technologies, in desert, forest, and plains cultures. But all illustrate that by taking up technologies, humans left the non-technological Garden to inherit the Earth. The price for that inheritance is to have taken up a technology. The point here, however, is to take measure. If the variants of human-technology relations cited here may be considered minimalist, there are close family resemblances to humans and their technologies.

To take measure is to interpret. Humans are self-interpreters, but not necessarily in terms of a self-enclosed self reference. They more often take their measurements in relation to other realms of being. For example, the animal kingdom almost always plays a role within this process. Our own biblical Eden myth, for example, clearly places humans over all animals and judges the distance between humans and animals to be fairly wide. The first command of Genesis is: "And God said, 'Let us make man in our image, after our likeness; and let them have dominion over the fish of the sea, and over the fowl of the air,

and over the cattle, and over all the earth, and over every creeping
thing that creepeth upon the earth. . . . ' And God said unto them, 'Be
fruitful and multiply, and replenish the earth, and subdue it.' " Nor have
humans been slack in following this instruction! As a dominant belief
within our own culture, this close relation to God and the more distant
one to the animal world has prevailed.

There is now a certain amount of re-examination occurring that
sometimes suggests a diminishment of distance between us and our
nearer animal relatives. While the purpose here is not yet to judge that
re-evaluation, the emergence of what I shall call proto-technologies
within the animal kingdom is also heuristically instructive for the task
of taking measure.

Jane Goodall, in her studies of chimpanzees, has documented that
they use and shape sticks to extract termites from their hills, they occasionally use clubs to kill small antelope, and they frequently pluck a
branch with a large leaf to shield themselves from rainfall. All such tool-
use behavior is clearly on the same trajectory as the human use of
tools, even though the objects used are rarely further shaped or carried
with the user.

Even lower on the scale, thorn-using finches discovered by Darwin in the Galapagos Islands, the sophisticated temperature-control
systems of social insects such as termites, and the regularly practiced
"agricultural" techniques of ants and other social insects all suggest a
kind of proto-technological attainment. These near relations to a technological culture may be noted without in any way diminishing the extent or level of development which occurs fully only within the human
domains on our planet.

The myth of the New Eden places any technological gestalt in
sharp focus in relation to the various degrees of nearness to a high
technological form of life. If animals have certain proto-technologies,
and if technologically minimalist cultures such as the Tasaday are far
relations to our maxi-technological culture, they nevertheless are already far from the abstract purity of the non-technological Garden.

But have we ever fully left that Garden? Do we have "memories"
which take the shape of limited events or occasions in which we experience the naked face-to-face of our primal pair? Who does not recall
or occasionally indulge in:
• the "skinny dip" of childhood.
• intimate sexual relations, perhaps on a moss bed in the forest.
• a plunge into the snow after the hot bath in a mineral spring.
• walking barefoot under the moon on a deserted beach.
Here are immediate bodily perceptual experiences at least temporarily
absent from the technological mediations which are more normal.
These are *direct bodily* and perceptual experiences of others and the
immediate environment.

I began with a deliberate listing of pleasant face-to-face occasions at the deliberate risk of suggesting the romanticizing of the face-to-face. Were the list to be left here, the nostalgia for innocence would become rampant. The list needs to be counterbalanced by equally negative examples of the face-to-face:

• I remember once, in the heat of a Kansas summer, scooping oats, only to have the wind blow the chaff over my unshirted chest, causing painful itching. A variation in which one rolls naked in the oats would non-technologically establish this itchy-being-in-the-world.

• Taking any of our warm and cold variants to the extreme—a fall into a boiling geyser or the icy water of a near-frozen lake—brings about negative, even deadly, variations of the naked face-to-face.

The full range of the face-to-face may be either positive or negative. What is essential is to isolate the direct, non-technologically-mediated dimension of those experiences. That dimension will be necessary for contrast with precisely those experiences which are technologically mediated.

The isolation of a *difference* between technologically mediated and non-technological experiences of the world will shape the initial part of the inquiry. The problem always encountered at this point is that such a difference almost always gets crossed with either a romantic or an anti-romantic interpretation in most examinations of technology vis-à-vis human life.

The yearning for innocence that dominates most romantic interpretations is very old. The Rousseauean tradition of the "noble savage" had its variants which stretched even into the precincts of both early exploration and certain anthropological interpretations. Whether coming of age in Samoa was ever as free of adolescent tension as once portrayed by Margaret Mead, we now know that the history of many Pacific Islanders had its own bloodier sides. They were apparently as prone to head-bashing as any, and intertribal wars may have been motivations for the much earlier exploration and settlement of virtually all inhabitable islands in the Pacific a millennium before Leif Ericson got to Greenland. Nor is the now-agreed-upon fact that up to two-thirds of live births were "birth controlled" at times by infanticide likely to encourage a romantic interpretation of what once was a more standard belief about Pacific paradises.

In more recent years—not long ago—the same search for innocence was applied to our near relatives, the primates. While the gorilla may remain basically vegetarian in diet, Goodall's studies of chimps disillusioned another precinct of hope for better behavior. She discovered that not only do chimpanzees occasionally eat meat (even using clubs to kill deer) but also that they occasionally kill each other, particularly the young.

But the point is not to go to the other end of the spectrum and argue for some new form of original sin. In the same debates concerning the innocence or non-innocence of either our human peers or our

near-human primate relatives, the new believers in original sin usually
espouse some variant of unchangeable human *nature* whose patterns
may be anticipated sociobiologically within some domain of animal
behavior; but the variety of behaviors among animals is too diverse to
do more than provide some field for a preferred selectivity, which in
turn is often used to reinforce a secret religio-political stance taken by
the interpreter.

What is needed is a much more radically demythologized story of
the structures and limits of human-technology and of the non-technological possibilities of relation to an environment, or "world." If this
essentially descriptive evaluation is taken into account, the tale of the
Garden may again be used as a limit-idea to delimit some of those aspects of the human experience which remain in some sense face-to-
face with others and the world.

Yet there is a second, limited sense in which we have never left
the mythical Garden. There are different degrees in which our experience of the world is not technologically mediated, at least at the center of our perceptual and bodily experience of that world. For
example, even though clothed and inside some "machine for living,"
as the functionalists have termed buildings, the normally sighted and
hearing person simply hears and sees what is immediate. In contrast,
anyone using corrective lenses or a hearing aid clearly has sight and
hearing mediated through a technology. At the level of touch, first in a
surface dimension, we always can become specifically aware of the
bodily surrounding immediacy of what we touch. In short, our sensory
life, even if at close range or enclosed, retains that sense of direct
perceivability and of bodily motility in the immediate environment.
How this bodily actional and perceivable experience differs from more
specifically technologically mediated experience will play an essential
role in the *initial difference* I am trying to isolate for this analysis.

Once having located this central core of perceptual, bodily experience of an environment, it is possible to point to both its *constancy*
and its *pervasiveness*. As long as I experience at all, I do so in bodily-
perceptual ways, and this is the case *inside* any technologies I may occupy. In a cold environment, I could tactilely experience the wind and
chill; but if I have "chosen" to mediate that cold by wearing down
clothing, I now substitute feeling the wind for feeling the warmth of
what I am wearing. In this case, the "environment" is simply brought
close and itself has the texture of one of the many cocoons humans
employ in all non-Garden situations. The technology (clothing), however, transforms this immediately experienced environment; and it is
that transformation which must be investigated.

Direct bodily-perceptual contact with an environment counts
as one side of the non-technologically/technologically mediated human experience that forms the focus of an entry into the analysis of
human-technology relations.

If direct bodily and perceptual contact with an environment, however limited, is constant, it retains the sense of our non-technological
garden existence. Yet care must be taken even here lest we think such
an imaginative variation simply captures the needed nuances to make
a sharp contrast between a technological and a non-technological form
of life.

The cultural or, better, cultural-technological forms of life which
circumscribe all our empirical human societies are also contextual in
terms of holistic gestalts. Virtually all human activities implicate material culture, and this in turn forms the context for our larger perceptions. Take, as an extreme case, the phenomenon of death:

The wildebeest would most likely be understood as a non-technological being, at least in its native habitat. Lions in this region are constant predators and frequently capture young calves. The mother
wildebeest, sometimes in concert with others of its kin, sometimes
alone, defends the calf until the lion makes the capture; but then, with
rarely more than a moment of delay, the mother wanders off and is
soon grazing again with the herd. Without deep speculation about
what and how the cow perceives its calf's death, something of a fatal-
istic or accepting aura surrounds this behavior. Contrarily, the chim-
panzee, as observers have noted, may carry around a dead infant for
days and even go into a rather obvious form of animal depression or
psychosis over the death; but in the end, the infant is simply dropped
and life goes on.

In the human case, however, there are burial practices, all of
which, in terms of material culture, involve technologies. Until recent
times, moreover, it was a cross-cultural commonplace not only to have
some practice with respect to the corpse but also to include in the
burial place a range of ordinary or represented artifacts and even other
live beings or their representations.

The twentieth-century discovery of certain ancient Chinese imperial burial chambers revealed entire armies (in ceramic) represented
along with the more traditional eating vessels, household items, pets,
and the like. Neanderthal burial places often contained clothing,
weapons, pots, etc. Death, as materialized in burial practice, may well
have been considered a passage; but far from its being a passage from
the totally familiar (worldly) to the totally unfamiliar (non-worldly or
supraworldly), the practices seem to evidence a kind of a material
continuity.

What is taken on the journey is what is used in daily life and what
falls within the circle of important experience. This includes artifacts,
particularly those which express the power of the individual (weapons
are common, but so is jewelry, which makes grave robberies so tempt-
ing). Sometimes even close relations or their substitutes were buried.
There were included, either in person or in representation, *other people* who were the companions or wives or servants of the dead. Live

burial of slaves (Egypt), sacrifices of slaves (Maya), co-burning of wives (India), and sacrifices of slave maidens (Viking) was widespread, as were the representations of concubines (Egypt) for the voyage. My point here is not to show the extreme male chauvinism in the past (even if the examples may serve that purpose) but to illustrate what might be called a variable border on what was considered within the power or possessive power of the individual. In societies where individuals are considered reduced to or restricted to autonomous selves, burial practices seldom include more than a few private personal possessions. It is, as it were, a difference in the effective "body" of the individual that is exemplified. The physical body remains only the center of some larger, radiating extent, which includes within its parameters the favored artifacts and living beings that belong immediately to it. One final comment before taking a less morbid example: The burials indicated above are of high-ranking, powerful figures within the societies noted. Commoners rarely had the same elaborate funerals. Class distinctions are apparently sometimes also indicated by burial practice; for example, it is now known that burial of individuals in a fetal position among Celts was not necessarily a sign of some more ancient practice. It sometimes occurred contemporaneously with full-length burial practices, and differentiated the status of the individuals. This difference, while clearly "economic" in the contemporary sense, was perhaps more indicative of the extent of the extended "body politic" of the individuals so treated.

In contrast to the non-human animal forms of the behavior toward death, human burial practices implicate aspects of human-technology relations. While I shall suggest that the experience of and relation to artifacts displays no clear and clean line of demarcation between human and animal; but, in fact, the essential ambiguity of technology is evidenced in the full range of human activity, including burial practice.

If human activity transforms the phenomenon of burial practice and in the transformation includes its technologies, a happier illustration may be found in culinary technologies and the phenomenon of eating. Claude Lévi-Strauss, during the height of his popularity as an anthropological theorist, made a strong nature/culture distinction a mark of his theory. A favorite general distinction along that line was the difference between the *raw* and the *cooked* (the former still standing under the aura of "nature"; the latter, clearly "cultural").

Techniques (even if not immediately implicating technologies) of food preparation, in contrast to simply eating food material as found, characterize human activity. The knowledge of which plants are edible and which are not, let alone the vast indigenous familiarity with therapeutic or poisonous plants, is truly astonishing among those cultures close to our mythical Garden state. (This knowledge is almost always *lost* by later forms of cultural life that no longer have a hunting and gathering or agricultural praxis. Here is a preliminary example of a counter-progressive state in "advanced" cultures.) This knowledge is complicated by knowing techniques to detoxify many toxic plants. The latter practice usually employs technological means (usually cooking, which uses fire, pots, etc.). Such techniques and technologies become increasingly complex with civilizational development. From a simple hand-and-finger eating to spoons to the full array of eating technologies (knife/fork/spoon multiplied by however many courses, to chopsticks), the variants are many.

Culinary technological gestalts are also interesting. The wok and steamer technologies of the East are stylistically and technique-wise different from the ovens and paddles or pots and stew-making of Africa, which again are different from the pot-and-pan cooking processes of Europe. Here is an initial example of technological multistability.

These examples, related to any aspect of human action, could be multiplied. Virtually every area of praxis implicates a technology. From burial to birth to eating and working, the use of artifacts embedded in a patterned praxis demarcates the human within his or her world. To reverse the usual equation, the technological form of life is part and parcel of culture, just as culture in the human sense inevitably implies technologies.

By looking at technologies in this initially broadest sense, we can note that, in contrast to the non-technological Garden, *human activity from immemorial time and across the diversity of cultures has always been technologically embedded.* This is not yet to take account of minimalist compared to maximalist or of traditional compared to high-technology developments, which are also distributed differently across time and cultures. It is to have anticipated how thoroughly the many forms of human life are materially conditioned.

This detour into the Garden as an imaginative contrast to the actual world of human-technology interactions has been taken for heuristic and suggestive purposes. A "return" to such a Garden is likely neither desirable nor possible. Given the extent and magnitude of the human population of the earth, were we to lose our technological capacity tomorrow, the result of the ensuing crisis would approximate, perhaps more slowly and painfully, precisely the likely outcome of a nuclear war in human devastation. The survivors would be those who somehow found themselves in the few garden spots of the tropics noted in our first excursions from the Garden. In this sense, a non-technological end of the world is not terribly different in result from a technological one, even though the former is merely an imaginative possibility and the latter something stronger. We have left the Garden and inherited the Earth.

2. Technology and the Lifeworld

The notion of technology I have developed to this point is as broad as possible while still retaining an emphasis upon its materiality. In the earlier navigational metaphor, all that has been distinguished is land and sea. It is now time to indicate more precisely which methods of navigation will be used to fulfill the voyage.

The term "lifeworld" will be recognized by the philosophically educated as one used by Edmund Husserl in his *Crisis in European Science and Transcendental Phenomenology* (1936). While I shall not borrow this term and its meaning without serious modification, it does serve to locate the inquiry within the traditions of *phenomenology* and its related *hermeneutic* origins. These traditions, however are not simply identical. Phenomenology, in an initial and over-simple sense, may be characterized as a philosophical style that emphasizes a certain interpretation of human *experience* and that, in particular, concerns *perception* and *bodily* activity. Hermeneutics, on the other hand, arose out of the disciplines of textual interpretation and later a (Continental) type of language analysis. If Husserl remains the central figure in phenomenology, his younger colleague, Martin Heidegger, was the central developer of a phenomenologically oriented hermeneutics; yet he also could be called the founder of contemporary philosophy of technology. The connections between these philosophical strands are not immediately apparent. The inquiry here joins these strands into one that proposes to descriptively analyze *technology and the lifeworld*. Beginning with a phenomenology of human-technology relations and then moving on to a hermeneutics of technology-cultural embeddedness, the first program is one that, while drawing from the traditions mentioned, is not limited to their past forms.

There is a certain risk in openly beginning with a phenomenology of human-technology relations because, in its experientialist focus, phenomenology has long been misunderstood as a purely "subjective" analysis. While in recent North American philosophy of science there have been some clarifications of this misunderstanding, a certain prejudice remains within the dominant traditions.

If it is true that phenomenological philosophy turns to the analysis of human experience, its understanding of the idea of *experience* is

not that of either common-sense understandings or of the standard
misinterpretations of phenomenology as subjective. I propose to grasp
this problem directly by underlining several aspects of experience as it
is understood phenomenologically.

In psychology—the discipline one might intuitively think the most
likely to inquire into "experience"—the term has long been passé. Experience became thought of as, redundantly, *subjective* experience,
and its access was *introspection*. The concept "behavior" substituted
for what had been experience; behavior was thought to be external
and observable. Recently a version of experience—which has as one
of its dimensions the non–externally observable interiority—has returned in *conceptual* psychologies. But in another sense, experience
never in fact left the practice of psychology. It is implicit in the very
experimental situation.

In practice, the psychologist constructs an experiment in which a
"subject" undertakes some test, task, or whatever. The psychologist—
who is also a subject in some sense—*observes* the experiment, inter-
prets its results, etc. But what is the status of the observation? Is it not
experience in a strong sense? Now, admittedly, psychology is not a
self-reflexive discipline, nor are most of the sciences self-reflexive in a
thematic sense. Thus observation—itself an experience—is functionally taken for granted, although there are controls within the community of psychologists for criticism of experimental procedure and thus
implicitly for the observation-experience. But what is obvious is that
without the observation-experience, there could be no science of psychology. Nor can this observation-experience be allowed to be
"merely subjective," lest psychology itself become a type of relativism
(to each his or her own observations).

A preliminary phenomenological analysis of the experiment-situa-
tion may help clarify the implicit notion of experience that functions
within psychology. Phenomenologists contend that all experience is
experience of *something*. That is, experience is referential, and that
which is referred to or experienced is anything which can fill in the
blank. In the case of the psychologist observing an experiment, the
blank might be:

Psychologist ⟶ (subject undergoing task)

If the psychologist does not actually observe the experiment, it may
simply be the results which are experienced:

Psychologist ⟶ (results)

In both cases, the psychologist *experiences* the referent, whether in a
direct observation or a secondary observation (and the phenomenologist might well point out that the first is an informed perceptual task,
the second a more explicitly perceptual-hermeneutic task).

The psychologist's experience is necessarily "subjective" in some
sense; but it is also more than subjective—it has its own context of con-
straint, or "objectivity." Such constraints include peer review (intersub-

jectivity), experiment design (context constraints), and the like. The trained, expert observation is simply a particular type of experience, but one that is surely not simply "merely subjective." And the same structure of experience of _____ belongs to each of the sciences.

If psychology, for its own peculiar reasons, has decided to reduce its phenomena to their "external," or directly observable, dimensions, physics has been more sophisticated with respect to difficult-to-isolate phenomena. For example, the "behavior" of atoms and their constituents is not directly observable but must be made available through a technologically mediated (instrumental) observation situation. The bubble chamber, accelerators, electron- and computer-enhanced microscopes, all bring into mediated or indirect presence the microphenomena which are of interest to the physicist. But the physicist, like the psychologist, would insist that the experience involved in observation must be of a special type. It must be informed and trained observation, and the observer must undergo an appropriate apprenticeship. Yet all observation remains experiential, without which there would be no science. Philosophy, and in particular phenomenology, takes in turn as one of its primary phenomena the *structure* of that experience itself. But if the structure of experience becomes a dominant theme of phenomenology, it does so because the full range and multiple dimensions of that structure must be examined. For this reason Husserl early characterized his initial steps as a "science of experience." Such a science cannot simply reduce its field to some arbitrary aspect of the whole. Both "external" and "internal," "subjective" and "objective" aspects must be included. In this sense, phenomenology retains a non-reductive strategy with respect to its field of inquiry.

In a deeper sense, however, phenomenology with regard to experience does not limit itself to a parallelism with psychology. Rather, it goes on to make much stronger *ontological* claims. This was pointedly so in the subsequent development of phenomenology by Martin Heidegger in *Being and Time* (1927). His existentialization of Husserl's earlier science of experience was, in fact, the development of what I shall call a *relativistic ontology of human existence*. Here I would point out that a relativistic account is not necessarily a *relativism*. Rather, a relativistic account is an account of *relations*. In Heidegger's case, it was an account of the human-world relations which determine and outline the dimensions of human existence (*Dasein*). *Being and Time* was an account of human spatiality within the World, of human temporality within the World, and of the various structures and dimensions of human-world relations.

Indeed, I suggest that one metaphorical model for understanding phenomenology is precisely that of a *relativistic* science. A simple way of stating this model is to indicate that the *primitive* of the system (the smallest or simplest unit) is itself a set of relations:

I---relation---World

And while I shall make these relations more precise when the program concerning human-technology relations is begun, here a preliminary comparison to a well-known relativistic example can be indicated. It is a modification upon the example Einstein used to illustrate a relativistic observation.

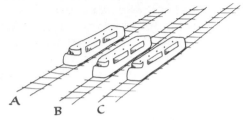

If there are three trains (see Figure 1) on three parallel sets of tracks, the motion observed will be relative to (a) the position of the observer in relation to (b) what is observed. What is relativistic about this account is that what is observed and the position from where it is observed and the interrelation between these two must be taken into account. Thus, if the observer were stationed in train B and noticed that train C was moving backward, several hypotheses are possible: either train B is stationary and train C is moving backward, or train C is stationary and train B is moving forward, making it appear to someone in train B that train C is moving backward—or both trains are moving. Adding observers in trains A and C and considering all train-movement possibilities shows the complexity of this situation, but in each case there is a "stability" within the observer-observed relation.

It also can be seen that such a relativistic account, which takes into its consideration *both* observer-observed (as a relationship), also can conceptually *absorb* any absolute or non-observational account. For example, were we naively to ask which train is "really" moving, we might construct a fourth position of observation, say, in a train-watching tower, D, which oversees the yard (see Figure 2). An observer here would not be subject to the relativity of the observers in the trains and could tell which of the hypotheses was "true."

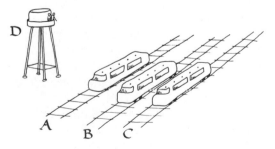

This privileged position does not escape the absorptive power of a relativistic account, however, because to take such a position as privi-

leged is simply to put in a different place the relativity between observer and what is observed. It makes the field of what is related more
comprehensive and complex; the field now consists of the trains as figures (moving or not) against a ground (which in this position is taken
as not moving), but it in no way eliminates the relativity of the account. The regional "arbitrariness" of giving superiority to the tower
position can itself be transcended, for were the earth itself observed
from a more distant point, it would be seen that what was taken for
the motionlessness of the ground itself is relative to the motion of the
entire earth through its orbit, and so on, ad infinitum. Yet the constant
of observer-observed would remain no matter what the distance or
position occupied would be.

A phenomenological account, like the relativistic one sketched, always takes as its primitive the relationality of the human experiencer
to the field of experience. In this sense, it is *rigorously* relativistic. The
relationality of human-world relationships is claimed by phenomenologists to be an ontological feature of all knowledge, all experience. Negatively, it would be claimed that there is no way to "get out of" this
relativistic situation, and any claim to the contrary can be shown to be
either naïve or misguided. In this sense, phenomenology is to all
foundationalist philosophies what relativity is to Newtonian physics.

These are general comments about a phenomenological model of
knowledge and experience. To drive us closer to the more specific inquiry into human-technology relations, a second science analogue may
be utilized. Whatever else may enter the analysis of human-technology relations, I wish to retain the sense of materiality which technologies imply. This materiality correlates with our bodily materiality, the
experience we have as *being* our bodies in an environment. The analogue science here would be animal ecology.

Ecology is also relativistic, but in a more concrete and biological
sense. It is the study of organisms *in relation to* their environments,
within some specific ecological system. In short, the organism is studied in relation to its field, however simple or complex. Here again is
also an analogue to a figure/ground model. The figure (organism) inter-
relates with its ground (environment), and the study of this interaction
is ecology. Phenomenology, particularly with respect to its existentialization of bodily existence, is a kind of *philosophical ecology*. But it
is an ecology with one difference: The "organism" which is to be
studied is not and cannot be studied "from outside" or from above
because, in this case, *we are it*. The human ecology which is phenomenology is thus doubly, or existentially, relativistic.

In the Einsteinian example, imaginative positions could be arbitrarily assumed; but in some more fundamental sense, there is always one
position which a finite and embodied being *must* assume—the one it
is. Similarly, ecologically, there may be a wide variety of organism-en-
vironment structures, but there are conditions relating to the one,

however complex, which we occupy. And that structure necessarily
has its bodily correlates.

By utilizing the analogues of these two relativistic sciences, it can
be seen that a phenomenological philosophy, while not eliminating interior or seemingly "private" aspects of human experience, is in no
way limited to them. On the contrary, its limits are those imposed by a
relativistic context of relations (I-world) and further constrained by the
explicit recognition that, in this case, one pole of the relation is what I
am (within the relation).

It is also at this juncture that the possible appropriateness of a
phenomenological relativity makes contact with an account of human-
technology relations. I have repeatedly insisted that the materiality of
technologies be maintained—the concreteness of such "hardware" in
the broadest sense connects with the equal concreteness of our bodily
existence. In its history, phenomenology took an increasingly *existential* development, and the term "existential" in context refers to perceptual and bodily experience, to a kind of "phenomenological
materiality." Technologies, in this sense, are not only appropriate for
examination; they fall almost naturally into such a philosophical focus.

Technologies, insofar as they are artifactual (in a range from simple entities to whole complexes of systems), are developed, used, and
related to by humans in distinctive ways. Yet while there is both a certain need to classify technologies as objects (which is often the first focus of objectivistic accounts), what will be focused upon here will be
their set of human-technology relations, the relations which can best
be exemplified in the kind of relativistic account suggested.

There are some tactical advantages to just such an approach:

First, phenomenological relativity will avoid what is considered to
be two extremist results of much of the literature in the field. The one
extreme position is one which ultimately *reifies* technologies into Technology. One version of such a move, already mentioned, is the one
which absorbs technology into *technique*. Technique, in turn, becomes
a certain way of practice and thought, which is so general as not to be
able to differentiate between particular human-technology relations
and is thus prone to overly metaphysical claims.

In my use, there may be techniques with or without technologies.
A sexual "technique" is not in itself a technology (although, in a derived and secondary sense, if such a technique is modeled after some
mechanical process, there may be an interpretive connection between
the two). Equally, techniques may be closely related to technologies,
particularly regarding patterns of use.

The second advantage of a relativistic account is to overcome the
framework which debates about the presumed neutrality of technologies. Neutralist interpretations are invariably non-relativistic. They hold,
in effect, that technologies are things-in-themselves, isolated objects.
Such an interpretation stands at the extreme opposite end of the reifi-

cation position. Technologies-in-themselves are thought of as simply objects, like so many pieces of junk lying about. The gun of the bumper sticker clearly, by itself, does nothing; but in a relativistic account where the primitive unit is the human-technology relation, it becomes immediately obvious that the relation of human-gun (a human with a gun) to another object or another human is very different from the human without a gun. The human-gun relation transforms the situation from any similar situation of a human without a gun. At the levels of mega-technologies, it can be seen that the transformational effects will be similarly magnified.

The third tactical advantage is to preserve in the analysis something of the dynamic or actional sense that obtains in human-technology relations. Not only are technologies artifactual but they are used (as well as developed, discarded, etc.) in their normative role. And although the use may be immediate, distant, occasional, or delayed, the human-technology relation implies human *praxis* or action. As a philosophy, phenomenology itself belongs to that family of praxis philosophies arising out of Hegel, Marx, pragmatism, and, in a derived sense, existentialism. Humans are what they are in terms of the human-world relation, but this relation in existence is actional. Both bodily-perceptual involvements and the development of the notion of praxis will be essential to the inquiry.

There is one further dimension of a phenomenological account that must be mentioned in this preliminary setting. Not only is such an account relativistic in the senses prescribed but it is also *structural*. In the examination of whatever range of relations the phenomenologist undertakes, what is sought is an account, an understanding of the structures of those relations. This will be the aim whether the structures turn out to be simple and monodimensioned or complex and multidimensioned. The examination should reveal both variant and invariant aspects to such structures. The question here, of course, will be: What relational structures obtain with respect to human-technology relations?

How are these different elements of a phenomenological account to be synthesized? How is the dynamics of perceptual-bodily activity in actional praxis to be combined with the elucidation of relational structures? That is the function of the notion of the *lifeworld*. Its origin, historically, is from Husserl in the later days of his career, but it also may have been his response to the earlier, more historical and existential account of human-world relations by Heidegger. In any case, in Husserl's development of the idea of a lifeworld, there lie suggestions that combine the elements needed to unify the theme of this inquiry.

On the surface it would appear that Husserl's aim in his own context is contrary to the focus needed here. The crisis in European science and philosophy that Husserl was responding to was, in part, the same as later was to face Positivism, a threat to the idea of rational and scientific progress. In his own answer, Husserl sought to develop a theory of rational progress through isolating the ways in which eidetic sciences—of which mathematics and geometry were the prime examples—could arise from previously more material or concrete contexts. Then such eidetic sciences could become idealized or autonomous from the material conditions and thus, in carefully qualified ways, be said to be cultural accumulations. His analysis thus focused upon the development of increasing degrees of ideality or abstraction.

In spite of this, there is a suggestive core contained in his derivation and concept of the lifeworld that can be redirected for the inquiry into *technology*. I shall later more fully discuss this derivation, but initially the illustration he gives in the famous appendix to the *Crisis*, "The Origin of Geometry," may open the way to the key aspects of the lifeworld notion:

> It is clear that, in the life of practical needs, certain particularizations of shape stood out and that a technical praxis always aimed at the production of particular preferred shapes and the improvement of them according to certain directions of gradualness. First to be singled out from the thing-shapes are surfaces—more or less "smooth," more or less perfect surfaces; edges more or less rough, or fairly "even"; in other words, more or less pure lines, angles, more or less perfect points; then, again, among the lines, for example, straight lines are especially preferred; and among the surfaces, the even surfaces, for example—for practical purposes—boards limited by even surfaces, straight lines and points are preferred, whereas totally curved surfaces are undesirable for many kinds of practical interests. Thus, the production of even surfaces and their perfection (polishing) always plays its role in praxis.[1]

What Husserl is interested in is the origin of geometrizing thought in practical activity. Lifeworld combines both a genetic (historical) account and a structural one. And the derivation clearly shows—in spite of Husserl's rationalism—a proximity to the praxis philosophies (such as those of Marx and Dewey) insofar as some species of action underlies later, more abstract developments.

In the example given, which I shall call a "furniture example," Husserl betrays his usual choice of familiar things as well as his implicit metaphysical framework. At the level of practical activity, things are material things and basically experienced as *plena* (multidimensioned and complex), and here they are acted upon as by a carpenter. The nascent geometer lurking in the carpenter begins to be born by selecting and preferring some abstract aspect of his materials (boards, as Husserl suggests). Here we must note that Husserl is quite critically aware that such a selection is, in fact, a choice and implies that other

---

1. Edmund Husserl, *The Crisis in European Sciences and Transcendental Phenomenology*, trans. David Carr (Evanston: Northwestern University Press, 1970), p. 375.

trajectories could be taken. The geometer's choice, however, origi-
nates in the preference for a set of shapes (points, straight lines, surfaces). And by perfecting and concentrating upon these, a new (and
non-carpenter) praxis may develop. (Husserl's examples always have
struck me as "cooked." Were the carpenter to be building a ship, for
example, his choices of shapes would be very different; they would include a preference for curves, compound curves, conic sections rather
than the simple angles, surfaces, and lines in Husserl's example. But
that choice would have made the connection with the origin of geometry more difficult indeed, since, in contrast to plane geometry, the
more complex geometry needed in the ship-building example could
hardly be the initial geometry. Historically, of course, ship building
continued without a geometrical science until quite recent history.)

A geometrical praxis, once derived and become autonomous, carries with it a new set of ways of seeing. It is a cultural acquisition that
can be repeated and that, once sedimented in cultural experience, becomes intuitive or taken for granted. (Here, another gross misunderstanding of phenomenology might be mentioned. The claim that
phenomenology falls into the "myth of the given," as Wilfred Sellars
might have it, is simply false. Husserl often referred to what was intuitively given, but it was used indexically; that is, it was to serve as a
marker to be investigated. Contrarily, *intuitions*, as the geometry example indicates, are *constituted*. Only when already constituted and
made familiar do they become fully intuitional, or "given.")

The problem that emerged for Husserl was that there now
seemed to be two levels of praxes: one material and practical (the car-
pentry context) and the other ideal and theoretical (the geometry context). Both belong in some way to the lifeworld, for both can be
familiarized within some praxical pattern. In part, however, this tension
arises because Husserl has combined two quite different praxes and
made one derivative from the other.

The modification I shall make here upon Husserl's example is by
way of distinguishing two senses of *perception*. What is usually taken
as sensory perception (what is immediate and focused bodily in actual
seeing, hearing, etc.), I shall call microperception. But there is also
what might be called a cultural, or hermeneutic, perception, which I
shall call macroperception. Both belong equally to the lifeworld. And
both dimensions of perception are closely linked and intertwined.
There is no microperception (sensory-bodily) without its location
within a field of macroperception and no macroperception without its
microperceptual foci.

The relation between micro- and macroperception is not one of
derivation; rather, it is more like that of figure-to-ground in that
microperception occurs within its hermeneutic-cultural context; but all
such contexts find their fulfillment only within the range of
microperceptual possibility. The implication for the notion of lifeworld

will be that this inquiry will be nonfoundationalist with some clear qualifications thus necessarily placed upon the idea of accumulations.

This modification upon traditional Husserlian distinctions should allow a clearer account of how lifeworlds change. Insofar as macroperceptions vary radically, the relation to any microperception must also vary in at least meaning context. Cultural histories are variant histories—but they remain focused in our bodily existences.

Most immediately, this modification calls for a double-sided analysis of the range of human-technology relations within the limits of microperceptual and bodily experience; the other side of the analysis must remain that of a cultural hermeneutics that situates our existential life. Insofar as bodily perception has a structure—I shall contend that its multidimensioned structure provides certain constraints upon perceivability—it also has a multistable range of ambiguity such that this structure is compatible with a wide range of different cultural-hermeneutic contexts. Thus the analysis must be completed by taking account of the latter field within which microperception can occur.

Virtually every example of a human-technology interchange can illustrate this interrelation. Take the following gloss upon the fox and grapes story: The fox, seeing grapes too high to reach by his bodily jumping capacity, concludes that the grapes were sour; but the human, at first also unable to reach or jump to the grapes, picks up a stick and knocks the grapes down, thus not finding it necessary to conclude that the grapes are sour. Both fox and human, in the most narrow microperceptual sense, perceive the grapes as edible and desirable, but the primitive technological context made possible by the stick changes the perceptual sense of grapes as attainable and, with it, the macroperception the human may have both of the object of perception and of his or her ability to attain that object.

3. Lifeworld: Praxis and Perception

From an actual human point of view, a lifeworld without technology must be at best an imaginative projection. Yet any analysis of such a world or even of the difference between technological and nontechnological worlds must begin in such a way that technologies may take their proper place in that world. The difficulties of not prejudicing the case in either direction are not without anticipations in the philosophical traditions I have adopted: phenomenology and hermeneutics. Before turning to my own examples I shall briefly look at three prototypical analyses within this tradition: Heidegger's hammer, Husserl's Galileo, and Merleau-Ponty's feather.

A. HEIDEGGER'S HAMMER

Martin Heidegger surely is known as one of the foremost philosophers of technology. As early as *Being and Time* (1927), there were prototypical analyses of technological experience and its implications. *Being and Time* actually preceded Husserl's *Crisis* (1936) and, as a response to and critique of his older colleague's work, it also helped provoke a turn in the latter's emphasis concerning the notion of a lifeworld.

What was at stake was a difference between epistemological and praxical orientations for philosophy. One way of characterizing the debate is to compare the ways humans know or interact with objects of knowledge to their interaction with use-objects or equipment. Heidegger held that the latter are closer to our ordinary experience and also that such experiences are more basic, with specifically thematized forms of knowledge of objects derived from the former. "The kind of dealing which is closest to us is as we have shown, not a bare perceptual cognition, but rather that kind of concern which manipulates things and puts them to use; and this has its own kind of 'knowledge.'"[1]

It was to this kind of knowledge that the famed hammer example was fitted. The hammer analysis was a typically Heideggerian workshop

---

1. Martin Heidegger, *Being and Time*, trans. John Macquarrie and Edward Robinson (New York: Harper and Row, 1962), p. 95.

example, quite likely taken from a shoemaker context. The analysis was to be a deep phenomenology of the actions in which hammering occurs in such a workshop. The primary variants of the example could be termed normal use—the cobbler or the carpenter using a hammer to drive nails—in contrast to an abnormal situation in which the hammer turns out to be broken or missing. Between these two variants, Heidegger performs his phenomenological operations. In this analysis, Heidegger developed his own version of a phenomenology, molded to a pragmatic emphasis. Taking tools or equipment as basic things encountered in a surrounding environment, he applied what we have already seen as the gestalt features of phenomenology: First, all objects (in this case, use-objects as well) are relative to a context. There are no objects-in-themselves. In a use-object context, "taken strictly, there 'is' no such thing as an equipment. To the Being of any equipment there always belongs a totality of equipment in which it can be this equipment that it is."[2] This field within which a tool is what it can be is a complex one filled with "involvements" or cross-relations. Second, these cross-relations have what might be called an instrumental "intentionality," or reference, defined by the work project. Equipment is "in order to _____. In the 'in-order-to' as a structure, there lies an assignment or reference of something to something."[3]

Even the simple example of a hammer fulfills these descriptions: the carpenter takes up the hammer—along with an apron full of nails, perhaps a handful of shingles—and pounds the nails through the shingles onto a roof. Here the involvement of the hammer with nails, shingles, and roof within the project is clear enough. And the focal reference is also apparent.

But in such praxical contexts, a third feature complicates the analysis, because there occurs a peculiarity with respect to the user. The tool or equipment—in use—becomes the *means*, not the object, of the experience. "The peculiarity of what is proximally ready-to-hand is that in its readiness-to-hand, it must, as it were, withdraw in order to be ready-to-hand quite authentically."[4] This withdrawal of the technology from within direct experience is what I shall later term an embodiment relation.

Even if objects of knowledge were "simply there" for inspection, in contrast to this indirectness of use-objects, the question of how to make use situations visible would remain a problem. Heidegger's attempt to do this is done by what I shall call a "negative" turn within indirectness. The complex field of involvements and the referentiality of the use context may be shown through breakdown, malfunction, or an absent tool.

2. Ibid., p. 97.
3. Ibid., p. 97.
4. Ibid., p. 99.

When we concern ourselves with something, the entities which are most closely ready-to-hand may be met as something unusable, not properly adapted for the use we have decided upon. The tool turns out to be damaged or the material unsuitable. In each of these cases, equipment is here, ready-to-hand. We discover its unusability, however, not by looking at it and establishing its properties, but rather by the circumspection of the dealings in which we use it. When its unusability is thus discovered, equipment becomes conspicuous.[5]

I shall not follow out all the subtleties of Heidegger's analyses here, but I will make note of two features of the analyses which will return later as peculiar selectivities. First, Heidegger is well noted by his critics for his choice of "toolshop" examples. His positive examples of technologies often are simple items, such as the hammer here or, later, a turn signal on an automobile. Similarly, in the realm of positively evaluated articles, one may find peasant shoes, windmills, or Greek temples, while hydroelectric dams on the Rhine receive negative characterizations. More interesting, however, is the derivation of both objectness and objective knowledge of certain aspects of use objects through an indirect or negative turn. I shall later point to what I think is the outcome of this implicit coloring of human-technology relations.

This forewarning is in no way meant to show that Heidegger's analysis is incorrect. That a context of involvements related to some technology can be shown by breakdown or malfunction does seem clear. Indeed, in the first serious world shortage of fossil fuel, the 1973 gasoline shortage, the set of involvements of automobiles in a very vast and complex network of industrial involvements became both obvious and frightening. It even stimulated some more serious thinking about the need for alternative energy sources and motivated conservational practices to some degree. I shall contend, however, that such a turn is not the only way to make a context apparent.

Returning to the analysis, one reason why the phenomenon of use-objects and their particular type of "knowledge" can be overlooked, according to Heidegger, is precisely because in uses of such technologies the user is "absorbed." "Has Dasein itself, in the range of its concernful absorption in equipment ready-to-hand, a possibility of Being in which the worldhood of entities within-the-world with which it is concerned is, in a certain way, lit up for it, along with those entities themselves?"[6] It is this absorption and the consequent withdrawal of the equipment in use which, Heidegger believes, justifies the indirect route he takes.

Were we to let stand by themselves the insights concerning our absorption in work projects and the concurrent withdrawal of the experienced tool, the wider implications of such actional relations might

---

5. Ibid., p. 102.
6. Ibid., p. 102.

be missed. Ultimately, Heidegger argues even as early as *Being and Time* that the praxical "knowledge" implied in technology points to an interpretation of nature itself.

> Any work with which one concerns oneself is ready-to-hand not only in the domestic world of the workshop but also in the public world. Along with the public world, the environing Nature is discovered and is accessible to everyone. In roads, streets, bridges, buildings our concern discovers Nature as having some definite direction. A covered railway platform takes account of bad weather, and installation for public lighting takes account of the darkness . . . . In a clock, account is taken of some definite constellation in the world system . . . . When we make use of the clock-equipment which is proximally and inconspicuously ready-to-hand, the environing Nature is ready-to-hand along with it.[7]

In short, implicit in such technological contexts is a view of the world—a view which Heidegger would later argue implies seeing the entire world as a "resource well" for human usage.

This is also to say that such uses "reveal" a world. Here, well before Kuhn's account of scientific revolutions as ways of seeing, Heidegger was claiming that technologies imply a mode of seeing:

> Our concernful absorption in whatever work-world lies closest to us, has the function of discovering; and it is essential to this function that depending upon the way in which we are absorbed, those entities within-the-world which are brought along in the work and with it . . . remain discoverable in varying degrees of explicitness and with a varying circumspective penetration.[8]

Thus, within the hammer example lies, potentially, a full entry into a philosophy of technology.

B. HUSSERL'S GALILEO

Nine years after *Being and Time,* Edmund Husserl's *Crisis* was published. In this work the concept of a lifeworld, complete with a revised emphasis upon praxis and perception, emerged. In "The Origin of Geometry" I have already noted that this "later" Husserl turned to a kind of furniture example of deriving the science of geometry from more mundane activities such as measurements.

> Even if we know almost nothing about the historical surrounding world of the first geometers, this much is certain as an invariant essential structure that was a world of "things" including the human beings themselves as subjects of this world; that all things necessarily had to have a bodily

---

7. Ibid., p. 166–81.
8. Ibid., p. 101.

character. . . . Further, it is clear that in the life of practical needs, certain particularizations of shape stood out and that a technical praxis always aimed at the production of particular preferred shapes and the improvement of them according to certain directions of gradualness.[9]

To have originated geometry from daily practices seems to be a kind of concession to the earlier Heideggerian praxis emphasis, but this concession remained only partial and was to cause a different set of problems for Husserl's version of lifeworld. First, a lifeworld was clearly conceived of by Husserl to be the most basic (foundational) stratum of World and also the broadest in scope. Any secondary or specialized "world," such as that of any science, must somehow refer to this basic world. This also applied to his derivation of geometry: "The geometrical methodology of operatively determining some and finally all ideal shapes . . . points back to the methodology of determination by surveying and measuring in general, practiced first primitively and then as an art in the prescientific, intuitively given surrounding world."[10]

Husserl's version of praxis in this wider cultural world was clearly not Heidegger's actional kind. It was rather the intuited material, bodily, and perceptual world of objects. "In the intuitively given surrounding world . . . we experience 'bodies'—not geometrically-ideal bodies but precisely those bodies that we actually experience, with the content which is the actual content of experience."[11] What is prescientific is still immersed in a characterization as *perceived material bodies*. The formal and abstract sciences such as geometry thus must move away from this materiality to become what they are. Husserl saw that as a process of perfecting, "Out of the praxis of perfecting, of freely pressing toward the horizons of conceivable perfecting 'again and again,' limit-shapes emerge toward which the particular series of perfectings tend."[12]

This is to introduce at the outset a bifurcation between the prescientific lifeworld and a scientific "world." This bifurcation will make the process of cultural acquisition difficult for Husserl. The issue comes to a head in Husserl's critique of Galileo, the introducer of a mathematized geometry into early science.

Husserl's Galileo may be said to inherit and stand between two world dimensions. On the one hand, he lives, as all do, in the world of perception in its prescientific significations, among "bodies," including his own. On the other hand, Galileo inherits a culturally acquired special praxis of geometrical thinking which he turns to new use in his physics. All motion is henceforth to be mathematized and thought of

9. Husserl, Crisis, p. 375.
10. Ibid., p. 27.
11. Ibid., p. 25.
12. Ibid., p. 26.

abstractly, distinct from the sensory qualities that belong to the very bodies that are in motion and that still belong in some sense to the prescientific lifeworld.

The problem revolves around sensory perception. Shapes, already closely subsumed under ancient geometry, are only a part of the sensory world. In addition there are what Husserl calls plenary sense qualities (colors will do, for example). These do not easily fall under the geometric praxis. "The difficulty here lies in the fact that the material plena—the specific sense-qualities—which concretely fill out the spatio-temporal shape-aspects of the world of bodies *cannot*, in their own gradations, be *directly* treated as are the shapes themselves."[13]

Some means must be devised, then, to account for these plenary qualities *or else* one must realize that the geometrical method is only a special and limited praxis related to one aspect of the world. Galileo's invention, Husserl claims, is the development of an *indirect* means of mathematizing the plena. Galileo must find a means to *translate* plenary phenomena *into* spatial ones in order for them to become available for geometrical analysis. And that is what he does.

Conceptually, this move, well known in both Galileo and Descartes, is one that first denies to the objects their plenary qualities in the doctrine of primary and secondary qualities. Put baldly, the object-in-itself is purely a geometrical entity, a *res extensa*; its plenary qualities are not located in the *object* but in the *subject*. Colors are "subjective." But now, since we see a thing as both extended and colored, there must be some way to subsume color into geometrical analysis. And it is here that the *indirect* geometrization begins to take shape. There must be some index of regularity that, while not directly spatial, can be related to some "spatial" measurement *directly* or through a process of *translation*.

> Now with regard to the "indirect" mathematization of that aspect of the world which in itself has no mathematizable world-form: such mathematization is thinkable only in the sense that the specifically sensible qualities ("plena") that can be experienced in the intuited bodies are closely related in a quite peculiar and *regulated* way with the shapes that belong essentially to them.[14]

In this perspective, not yet intuitive, Galileo dramatically paves the way for modern physics such that today second thoughts are rarely given to the procedure:

> What we experienced in prescientific life as colors, tones, warmth, and weight belonging to the things themselves and experienced causally as a body's radiation of warmth which makes adjacent bodies warm, and the

---

13. Ibid., p. 33.
14. Ibid., p. 35.

like, indicates in terms of physics, of course, tone-vibrations, warmth-vibrations, i.e., *pure events in the world of shapes.*[15]

So much of this is taken for granted that undergraduates can even say that they "see" wave lengths.

This is to say that once gestalted, the Galilean perspective becomes a kind of macroperception that can be taken for granted with new modifications possible. It is precisely here that the ambiguity in Husserl reaches its own apex. For *if* the new means of understanding phenomena becomes a genuine cultural acquisition as the investigation into the origins by means of a praxis becomes transparent, it overlooks the contrast with the fundamental lifeworld.

> But now we must note something of the highest importance that occurred even as early as Galileo: the surreptitious substitution of the mathematically substructed world of idealities for the only real world, the one that is actually given through perception, that is ever experienced and experienceable—our everyday lifeworld. This substitution was promptly passed on to his successors, the physicists of all the succeeding centuries.[16]

Husserl seems to be saying that at base the lifeworld is and must be the sensory lifeworld, based in the relations between actional humans and the concrete, material world of things and beings that are bodily. And these are intuitively, perceptually available to everyone. Then, a second type of intuitability also occurs such as that exemplified in the Galilean revolution, one in which certain combinations of praxically attained perspectives can make possible another intuitable attainment, a cultural acquisition like a science.

Such an acquisition, however, is also ambiguous, because what is gained by the very means of mathematization, Husserl argues, loses an essential sense of concreteness by overlooking the fundamental lifeworld.

> Galileo, the discoverer . . . of physics, or physical nature, is at once a discovering and a concealing genius. He discovers mathematical nature, the methodical idea, he blazes the trail for the infinite number of physical discoveries and discoverers. [But] immediately with Galileo, then begins the surreptitious substitution of idealized nature for prescientifically intuited nature.[17]

The acquisition of the new paradigm conceals the fundament of the ordinary dimension of the lifeworld.

Husserl's Galileo thus stands caught between a prescientific but perceived lifeworld and a scientific but unperceived world of ideality. I

15. Ibid., p. 36; my italics.
16. Ibid., p. 49.
17. Ibid., pp. 50, 52.

have already suggested that this anomaly is, to my mind, caused by Husserl's own reconstruction of the origin of science. By making the material-sensory world both prescientific and foundational on the one hand and emptying the scientific world of perception and praxis as a "derived," although special, "world" on the other, Husserl misses the interrelation between what I am calling micro- and macroperception.

Such a Galileo is not unique to Husserl; he is the Galileo of most interpretations that privilege one version of theory over praxis. He is a Galileo immersed in hypothetical-deductive reasoning of the dominant view. Of course he is also the experimental Galileo, although Husserl ignores the praxical elements of this practice. This view of science concentrates upon what and how science "thinks" rather than upon what it "does."

Science and its mode of seeing is now part of our macroperceptual world. It is a cultural acquisition that situates the very bodily perception we retain, but it overlaps or does not overlap with what must be, for us, a lost prescientific world. What is equally missing from this analysis is the actual praxis of modern science from its beginnings—its embodiment in technologies, its instrumentation. For through instruments, science in its modern form never loses its perceptions. Whereas Heidegger implicitly begins a philosophy of technology in relation to these analyses, Husserl does not yet open the door to such a philosophy. We must return again to Galileo.

C. MERLEAU-PONTY'S FEATHER

The third prototypical analysis in this heritage comes from the work of Maurice Merleau-Ponty. His *Phenomenology of Perception* (1945) had the advantage of both Heidegger and Husserl as predecessors. There is a sense in which Merleau-Ponty became the perceptual aesthetician of phenomenology. Here one finds a descriptive richness much subtler than either the Heideggerian workshop or the Husserlian furniture polisher. Note the following:

> A wooden wheel placed on the ground is not, for sight, the same thing as a wheel bearing a load. A body at rest because no force is being exerted upon it is again for sight not the same thing as a body in which opposing forces are in equilibrium. . . . Vision is already inhabited by a significance which gives it a function in the spectacle of the world and in our existence. . . . The problem is to understand these strange relationships which are woven between the parts of the landscape, or between it and me as incarnate subject. . . . Sense experience is that vital communication with the world which makes it present as a familiar setting to our life. It is to it that the perceived object and the perceiving subject owe their thickness.[18]

18. Maurice Merleau-Ponty, Phenomenology of Perception, trans. Colin Smith (New York: The Humanities Press, 1962), pp. 52–53.

The question arises: Is this aesthetic frillery, or does it display the basis for a postmodern awareness of a subtle perception sufficient to both science and artistic sensitivities? For what is finally surpassed here are the bare bodies of Cartesian science and the bare sensations of empiricist psychology.

What Merleau-Ponty adds to this procession of prototypical analyses is a strict phenomenological correlation between a "lived" body and the perceived world. Lifeworld perception is perception which, microperceptually, always implicates the body-in-action:

> What counts for the orientation of the spectacle is not my body as it in fact is, as a thing in objective space, but as a system of possible actions, a virtual body with its phenomenal "place" defined by its task and situation. My body is wherever there is something to be done.[19]

Bodily existence is actional and oriented. And it is correlated with a surrounding world open to action:

> We must not wonder why being is orientated, why existence is spatial, why . . . the body's coexistence with the world magnetizes experience and induces a direction in it. The question could be asked only if the facts were fortuitous happenings to a subject and an object indifferent to space, whereas perceptual experience shows that they are presupposed in our primordial encounter with being; and that being is synonymous with being situated.[20]

Merleau-Pontean microperception is always a kinesthetic perception:

> My body is geared into the world when my perception presents me with a spectacle as varied and as clearly articulated as possible and when my motor intentions, as they unfold, receive the responses they expect from the world.[21]

And it is out of this sensitivity towards a body-world correlation that Merleau-Ponty anticipates a different role for technologies within the realm of perception and praxis. The lived or virtual body as an experienced bodily spatiality can be "extendible" through artifacts:

> A woman may, without any calculation, keep a safe distance between the feather in her hat and things which might break it off. She feels where the feather is just as we feel where our hand is. If I am in the habit of driving a car, I enter a narrow opening and see that I can "get through" without comparing the width of the opening with that of the wings, just as I go

---

19. Ibid., p. 250.
20. Ibid., p. 252.
21. Ibid., p. 250.

through a doorway without checking the width of the doorway against that of my body.[22]

Here again, in what I shall term embodiment relations, Merleau-Ponty discerns that perception may be materially extended through the "body" of an artifact. Perceptual extension is not limited by the outline of my body or the surface of my skin. The description becomes more precise in the "blind man's cane":

> The blind man's stick has ceased to be an object for him and is no longer perceived for itself; its point has become an area of sensitivity, extending the scope and active radius of touch and providing a parallel to sight. In the exploration of things, the length of the stick does not enter expressly as a middle term: the blind man is rather aware of it through the position of objects than of the position of objects through it. The position of things is immediately given through the extent of the reach which carries him to it, which comprises, besides the arm's reach, the stick's range of action.[23]

Here is a basis for perception at a distance, mediated through an artifact, a technology; and here is a latent phenomenology of instrumentation. One can see that this analysis complements Heidegger. In the hammer example, the tool "withdraws"; but in the Merleau-Pontean feather or cane, it is a part of the world which is reached through this withdrawal.

A second factor fitting into Merleau-Ponty's refined perceptualist scheme is the interrelation between sensory perception and the role of culture. In his later work *The Visible and the Invisible* (1961), he took explicit account of "perceived" culture:

> It is a remarkable fact that the uninstructed have no awareness of perspective and that it took a long time and much reflection for men to become aware of the perspectival deformation of objects. . . . I say that the Renaissance perspective is a cultural fact, that perception itself is polymorphic, and that if it became Euclidean, this is because it allows itself to be oriented by the system. . . . What I maintain is that: there is an informing of perception by culture which enables us to say that culture is perceived.[24]

Relating this observation to the two different interpretations of science by Heidegger and Husserl, one can see that science—for Merleau-Ponty—could be such a perceived culture. Here is a kind of macroperception which informs or orients bodily perception itself.

Those three phenomenologies point in distinctive ways towards a

---

22. Ibid., p. 143.
23. Ibid., p. 143.
24. Maurice Merleau-Ponty, *The Visible and the Invisible*, trans. Alphonso Lingis (Evanston: Northwestern University Press, 1968), p. 212.

phenomenology of human-technology relations. None of these analyses in its historical setting was intended to fit into a philosophy of technology, but each points to elements that become crucial here; and whereas all phenomenology is in some deep sense "perceptualistic," the version of phenomenology I shall outline here draws from, but also differs from, those of my acknowledged tutors. What remains for the first program, a phenomenology of human-technology relations, is to reinsert the role of technologies in all the dimensions of the lifeworld.

4. Adam and Galileo

Imaginatively return to the New Eden and our pretechnological New Adam. What would our naked Adam see with his naked eyes were he to gaze upon the nighttime skies? From our present situation we know that this question is doubly suspect. Not only must we ask it from the very midst of a technologically constituted world but also we must suspect that the very imagination of an "innocent" perception is at least an abstraction.

The nighttime sky is a virtual paradigmatic perceptual ambiguity. Were we to imagine the naked vision of Adam, we might be tempted to characterize this perception as a dark expanse, a blackish perceptual field with points of light in varying degrees of intensity and random patterns. If our Garden is in the Northern Hemisphere, there would be a certain whitish, but amorphous, area spanning the middle. We might go on to characterize this gestalt as a microperceptual base to which could be added macroperceptual interpretations. Our pretechnological Adam might see such a microperceptual vision.

But as soon as we begin our variations from within our knowledge-saturated age, even at the level of minimal television-mediated awareness of science documentaries, we would suspect that there are no such Adams. Instead, in every case of a prescientific, even if not pretechnological, culture, we find any vision of the heavens much more richly described. Indeed, the bareness of our presumed microperception would probably not be recognized by our Adam at all. Were he somehow critically acute, he would point out that such language clearly belongs in a unique way to our own tribe— for surely no one sees the heavens in such a reductionistic way!

What we take as a naked vision already contains the same secret relation to macroperception. The histories of perception teach us that every version of microperception is already situated within and never separate from the human and already cultural macroperception which contains it. There is no simple seeing; there is only situated seeing that is both a seeing *as* _____ and a seeing *from* _____ . Yet here are initial clues for a phenomenological examination.

Any empirical Adams would see the heavens as _____ , within some concrete macroperception. The ancient descriptive cosmologies such as those of the Middle East take our visual sky *as* a dome, black at night, blue in the day, but obviously "solid." There is no break in its being colored and in being such-and-such a distance from earth. The lights of this heavenly dome might be holes in the dome through which the extra-heavenly fires show, or we might find that the stars, sun, and moon are travelers along the dome. Contrarily, were we to study Far Eastern cosmologies, we might find that the sky is the *open*. It stretches outward indefinitely beyond vision. Both these descriptive versions are microperceptually possible.

Similarly, were we to take what we first described as random patterns (our macroperceptual variant) to our ancient empirical Adams, we would find that there are groupings which are also seen as _____ . There are the zodiacal figures of the Greeks and Romans, the dippers of our North American ancestors, or the whale, seal, and fish of the farther northern peoples. Each tribe groups stars into patterns, not randomly, not consistently with one another, yet always into groups. The groupings, moreover, reflect life patterns and actions of the grouping peoples. In an agricultural society, the star patterns related to the various cycles of the seasons and reproduction. If navigation was the activity, the stars would relate to destinations. Questions concerning the relative distances of stars from the earth, for example, were not only absent but would not make sense within the cultural paradigm.

At the microperceptual level, the absence of such questions makes its own sense. Objects at great distances are foreshortened. We do not (microperceptually) perceive relative distances among the stars, and that is perhaps one reason why the ancient cosmologies placed the stars much closer to earth than more recent ones. The amazing results the Greeks obtained, which anticipate modern measurements, arose both from their mathematical beliefs and from their primitive use of *measuring instruments* Their perceptions were already becoming partially technologically mediated, but the use of instrumentation remained highly restricted. There were no optical technologies at all.

The ancients observed keenly, dynamically. They not only knew the patterns that are "constellations" but also mapped the motions of stars and other heavenly bodies perspicaciously. We now know that they employed large and monumental "sidereal compasses" such as Stonehenge to make such mappings. These circles of stones were aligned to precisely locate positions indicating the exact times of the solstice or to the risings and settings of important stars or planets. The calendar technologies, sun-pinpointing observational tunnels, and other instruments for detailing motions were keyed in to the variant patterns of daily life such as the cycles of the seasons and astrological concerns. There were no bare perceptions, yet all their perceptions included the naked observations which were made to fit within the forms of life that stretch to the heavens from the earth.

What, now, is perceived as we turn to our own variation? Today's

precocious child can tell you that the heavens are made up of galaxies
of several shapes, spirals or concentric circles; that there are unseen
black holes and radio stars out there; that the stars are not all of one
size but include white dwarfs, red giants, ad infinitum. These descriptions, however, do not easily mesh with the ancient views of the
nighttime sky—they contrast with it. There is a distance between what
is seen looking up on a clear night away from the light pollution of urban areas and what is depicted for us in the dramatic photographs and
representations of nebulae and galaxies. There is no immediate way in
which our nighttime perception can accommodate this portrayal. And
here we might discover a "distance" between the older version of a
lifeworld of immediate perception and the scientific world which distances us from the first.

Were we to come to this conclusion, however, we would be leaving out precisely the factor which makes for the distance and the simultaneous closeness—technological instrumentation. That is because
there is an essential, *technologically embodied* difference between our
perceptions and those of any of the ancients. Our perceptions are not
naked, but mediated. We see by means of first optical and then radio,
spectrographic, and other technologically embodied visions (or hearing, or touch). While there is thus a seemingly great distance between
the naked viewing of the nighttime sky and the depiction of the heavens in contemporary astronomy, the distance itself is mediated
through our sensory technologies. That is what places us far from our
mythical Adam.

Both the distance and the proximity of technologically mediated
and naked vision are important. No more than the ancients can we
micro-perceive the distances between stars. Our nighttime looking,
unaided by optics, cannot directly tell us Rigel is closer or farther from
the earth than Polaris, only that one is brighter than the other. Their
relative distance cannot be perceived even through fairly powerful
telescopes. We discovered the clues to vast distance in part because
the apparent sizes of the disks of the stars remained the same in early
telescopes, whereas the moon and the wanderers—or planets—
quickly changed. Yet our picture, macroperceptively, is what it is
through the instrumentation that distances us from all Adams. Their
arrangements are not "wrong" or ours "right"; they are simply incommensurate. Both are limited to the relativities that bind the viewer
and the vision, although in instrument-mediated vision there are new
factors.

Not only have our perceptions changed—those embodied
through instrumentation are incommensurate with naked observation
in however small degrees—but so also have our praxes. There remain
only vestiges of the linkage of the ancient agricultural world to our vision. We suspect that the cycles of sunspots may be related in some
way to long-term periods of weather patterns and these, in turn, to the

cycles of wet and dry, but our instrumental inclinations enter into our own praxes in ways discontinuous with the worlds of the ancients. We seek the very borders, if there are any, of the universe. The search itself is made possible only through the development of more and more powerful instruments. This inclination not only is made possible by technologically embodied vision but arises within the context of instrumental embodiment. Our episteme is different from Adam's, but there remain *interstices of intersection.*

The interstices lie between the flexible and ambiguous ways that micro- and macroperceptions are linked. However complex and multidimensioned they may be, there are structural features discernible to a phenomenology of perception. Those to be isolated here will concern the differences between technologically mediated and naked perceptions. I shall begin by remaining within the domain of the familiar. This includes a customary use of visual examples, which are echoed in the very metaphors we have used to describe science as a "way of seeing."

The *difference* between New Adam and ourselves is that between mediated and non-mediated microperception. We do not need to return to the Garden; in spite of our own technologically textured world, we still have non-mediated perceptions within the microlevel: I stroll outdoors during the summer, down to the beach. There, disrobing except for a modest set of swimming trunks, I first sit on the sand and look about me. The tactile sense of the breeze, the warmth of the sand, the sound of the ripples from the waves, the vision of the Persian reeds bordering the beach, the fully leaved trees across the cove—all are present to my senses non-mediatedly. I shall call this simply a direct (micro-) perceptional situation. I take it for granted. It is familiar. And it is usually unexamined. It is a fundamental way in which I experience my immediate environment or surrounding world. Such direct or non-mediated perceptions I shall formalize as *I-world* relations.

These relations already include a multidimensional complexity. What I experience, anything in the world near me, is both *seen as* and *seen from.* Waves, trees, reeds, sand—all are the familiar "things" of experience, ordinary beings towards which my flexible senses may be directed. I do not experience anything like a building up of these immediate things—they are already there, already what they are. But I do experience them as within a panorama, a field display. The trees are across the cove, with the beech tree to the left of the sycamore maple. The seagull flies above the swimming swan who is directly before me. The panorama extends to the very fringes of what I see around me.

I experience differences between what stands out within the panorama in direct relation to my attention and direction of gaze. I am attracted, by curiosity, to the horseshoe crab that now makes its slow approach to the edge of the beach, followed by its insistent mate. When I am thus directing my sight, those items in the distance become less distinct and more background-like to the immediate drama near my feet.

I may not be explicitly aware of this directionality of my gaze nor of the gestalt relation between foreground focus and background field, but it can be easily recovered in reflection. Also, I can recover my dynamic seeing as simultaneously a seeing *from*. That immediate panorama *reflexively* locates me. What is before me reflexively points back to a being-*here* which is concrete, the bodily spatiality which *I am;* but this bodily space is not distinct. It, too, is flexible and multidimensional.

I experience my hereness as a bodily kind of ambiguity with its felt foregrounds and backgrounds as well. My step to the now sun-heated sand suddenly focuses my awareness on the interface between feet and sand, and I move quickly to the water to change the heat to cool. But now relieved, I return to being focally concentrated upon the mating crabs. I am immersed in the surrounding world, but this immersion is as flexible and dynamic as the panorama about me. This is the chiasm, the intertwining of the flesh of which Merleau-Ponty spoke in his last interpretations of perception.

All of this is the interstice I share with our imagined New Adam. What he does not share with me, however, is the mediated, the embodied perceptual experience I have when I leave the Garden. Not much of my life is lived nakedly; when it is so lived, it is never far from the material clothing that is our technological embodiment. In fact, my distant world presents itself to me in my middle years with a certain visual ambiguity I had not known in my youth. Even to the beach I now take my eyeglasses, together with my bathing suit, which preserves the propriety of civilization that calls for technologically enclothed modesty in Poquott.

Nor does our Adam experience the *difference* between his nakedly perceived world and ours. Here lies a clue to what in the New Eden would be the loss of innocence but, once out upon the earth, becomes more than the knowledge of good and evil. It is the knowledge born of differences, of variations which hold the secrets to a world both wider and less innocent than Adam's. But just as the very familiarity with our Adamic dimension may be left unexamined and undiscerned, so too can both the quality of our mediated relation to the world and its difference from our naked relation to the world. To examine both mediation and its difference is to enter the phenomenology.

The first set of variations is optical and may be placed upon what may be called an *optical continuum*. Imagine our Adam climbing a promontory in his garden kingdom and discovering the panorama of the jungle. His contemporary counterpart may do so in a different manner—for example, by taking the elevator to the upper floors of a skyscraper and viewing the urban skyline through a newly cleaned plate-glass window.

Here we have what might seem, at first examination, a minimal difference between a direct or naked view of the world and a technological (here optical) mediated view. Formally, we may note that the window occupies a position *between* the observer and what is observed. The I-world relation is changed to I-window-world. This is more than a formal change; the way world is experienced is changed *ontologically.*

This, however, may not be immediately apparent. Indeed, the uncritical viewer might well be tempted to say that the panorama is exactly the same through the plate glass–embodied vision and the naked vision, for what seems first to be striking is the very *transparency* of the medium. The glass allows me to see through it, to the world. The glass, recalling Heidegger's term, "withdraws" from the projected aim of vision. It could even be said to be taken into my seeing, as eyeglasses are; and were we to formalize the I-window-world relation more specifically, we might well diagram it as

(I-window)-world

Here the parentheses signify that the window withdraws to such a degree that it is, at best, the *means* and clearly not the *object* of my vision.

There is, in most situations of this type, a certain amount of backglare, possibly some barely discernible translucency residual in the best of plate glass, just enough that a careful variation between the two views reveals that the withdrawal of the window is never *total.* There is also a slight but discernible flattening of the panoramic depth. But, imaginatively vary these extremely subtle differences between an attempted (but never actually attained) *pure* transparency, and some greater degree of translucency as a continuum toward the opaque.

A total opacity would not allow our visual experience to be mediated beyond the glass. The plate glass itself would be changed from means to *object* of experience. It would be the terminus of the visual gaze. Only when it remains at least partially transparent can it be taken into our vision in such a way that we see through and beyond it. Between total opacity and the highest degree of transparency there lie many other possibilities, and these possibilities would all be more dramatic transformations upon any previously nakedly seen world. A tinted window transforms the coloration of the panorama, the polarized window transforms the light and shows relations within the panorama, etc. One can imagine here a range not unlike that of the varieties of painting styles in their variants upon the landscape. The analogy is not without import for the technologically embodied scientific world we inhabit. If by varying the coloration in shadow, by making shadows blue and green instead of gray (as did the impressionists in art), we learn to see differently. The same occurs in the contemporary scientific variations with "false color" or by infrared photography. We learn something new of the physical world through such differences, and there is "truth" in each variation.

Let us remain, however, along the trajectory that varies only within colorless transparency. Our window glass might contain small bulges and uneven surfaces, like the old window glass of New England houses, in which case our naked and mediated visions would vary in distortion. Then, somewhere in history—the eleventh century—someone discovered that certain bulges *magnified* what was seen, and the lens was invented. Magnification, once discovered, suggests a new *trajectory*. If a little magnification shows something to be "bigger," what would more magnification show? Here the examination must move slowly because there is much that can be missed by the very first blush of a *fascination* with technological possibilities.

Early lens use was slowly adapted to the most familiar use of today, spectacles (thirteenth century) or eyeglasses, which in the twentieth century have become contact and even implant lenses for greater embodiment. But as early as the thirteenth century there were not only spectacles but even compound lenses and a treatise on lenses (Vitellio, 1270). Eyeglasses, with what later would seem a relatively small magnification, bear a short initial consideration: First, they are even better examples of the (I-glass)-world relation than the plate glass. In use, we barely notice them; they "withdraw" and become the transparency by which we see the world. But, like Heidegger's hammer, they may become conspicuous by being absent, by having been broken, by no longer serving as the proper prescription—in all of which cases a functional obstinacy obtrudes. Then the meaning of eyeglasses changes from means to object of experience.

In the magnificational capacity of the eyeglass, there is a certain shape to its technological "intentionality." Magnification selects the panorama in a certain way, and in the process, there is a change of both time and space. My seeing *as* is a magnified seeing *as*. When a new prescription is taken up, one has to relearn the minor adjustments to the very way one embodies ones vision lenticularly—this could be experienced more dramatically were one to put on lenses much stronger than required. Here the variation from normal motor activity is much more marked.

This is because all our seeing *as* is also a seeing *from;* and the transformation of vision through lenses changes, however slightly, our sense of bodily space. What was farther is now nearer; and when motility is involved, this calls for a new adjustment. The most radical such variation is the well-known inverted glasses experiment in which the viewer sees the "world" upside down. It takes several days for the viewer to relearn his actional gestalt such that the world appears "right" again—but it does.

Physiology has difficulty accounting for this phenomenon. A more gestalt-oriented phenomenology recognizes that I *am* a certain constellation of relations to a world and that this constellation is learned and relearned dynamically. Right side up and upside down are relativities within the context of bodily motility, as Merleau-Ponty has already pointed out.

The bodily space of vision even applies to less motile situations and may be noticed in fixed observations. There is even a sense in which our usual description of what magnification does is wrong. We say the microscope makes the microscopic "bigger," but this is so only in a comparative relation between what we see with the naked eye and what is seen through the microscope. More phenomenologically put, what changes is the sense of distance and closeness to the object of vision. The paramecium seen through the lens is no bigger or smaller in my visual field than other things that I place close to my nose; but I, in a partially irreal bodily sense, am now "closer" to the paramecium. What I took as my "real" or naked bodily space is transformed through the microscope. This transformation of bodily spatiality is noticeable but less dramatic with eyeglasses.

Because eyeglasses are mostly worn to correct vision, to retranslate the blurred world into a clear and distinct one, we gladly accept them. But the change is not without price, for who would not prefer to see the world clearly without them? Simultaneously with the newly corrected world, there is a reduction of it. There is, for example, the very small price of caring for our mediation equipment itself: Now I have to care for my glasses or my contacts and the paraphernalia which go with them. But more, particularly with eyeglasses, my world now comes to me *enframed*, intruded upon, perhaps almost imperceptibly, with the backglares that occur, with the dust or water spots that accumulate, and with a fringe awareness that the world seen through the lens is, however slightly, changed. With contact lenses, which are even closer to my body, the situation is improved and such side effects as backglare and frame intrusion disappear but, again, at the price of the now tactile "intrusion" or touching of my body that contacts bring (bits of previously unnoticed dust now are tactilely magnified as irritations, etc.). Yet few people who need glasses would cease to pay this price. But *for every revealing transformation there is a simultaneously concealing transformation of the world, which is given through a technological mediation.*Technologies transform experience, however subtly, and that is one root of their <u>non-neutrality</u>.

It may now be suspected that if the trajectory towards greater magnification is followed—and it is latent in the very discovery of the lens—one may also expect to find a greater and more obvious magnification/reduction structure to embodied vision. By our standards, these early instruments were relatively poor ones; but they illustrate even better the magnification/reduction phenomenon.

To look through a more powerful lens is to achieve a greater magnification. The object seen appears closer, more detailed; my bodily position toward it is simultaneously placed within a greater sense of quasi-closeness. The object is also brought more focally into the center of my vision. Yet this closeness is simultaneously accompanied by other changes that reduce other aspects of the object seen. For example, the greater the magnification, the thinner the focal plane—the sense of depth is lost as magnification increases. So is the expanse and location of the object in its own environment. To see the moon through a telescope is to see it close up but also to lose it in its position in the sky. Lens technology transforms the very sense of space that I experience, in a significant modification of both bodily and world space. It transforms it into a kind of irreal, flattened, and narrowed "world." Its distance is always a peculiar kind of *near-distance*. But we "forget" this as we learn to embody the technology into our familiar actions.

In spite of this reduction, which is noticeable, we continue to follow the fascinating trajectory of the more dominant magnificational dimension of embodied vision. It is possible virtually to overlook the reduction itself, and here lies another secret to Adam's leaving the Garden: What the magnification shows, as it becomes stronger, is genuinely new—the undiscovered. To be visually embodied through lenses is to enter macro and micro "worlds" which were previously unknown, and to do so technologically.

I have been phenomenologically describing a technologically embodied vision, the vision of the world mediated through variations upon optics, simple and even primitive optics compared to the diversity of optics in the late twentieth century. Latent in our first movements from the Garden, there already lie different seeings of the world. Within the context of simple optical technologies, one can also relocate Husserl's paradigm figure in the sedimenting of modern science: Galileo.

In recent years there has been a plethora of books on Galileo, many of which debate the traditional issues about theory in relation to experimentation. I shall not enter these debates but rather take two other interpretive themes that correspond to the lifeworld and the perceptual emphases of this inquiry. Husserl perhaps first showed, pre-Kuhn, how Galileo made a new paradigm of science possible.

Husserl's Galileo discovered the indirect mathematization of the universe but in the process also originated the great forgetfulness of the lifeworld, which Husserl lays at the door of Galileo's revival of an ancient depreciation of the senses. Galileo, prior to Descartes, distinguished between primary qualities, qualities that were *in* or belonged to the object, and secondary ones, which were to be relocated *into* the subject or observer—thus making certain qualities "objective," others "subjective." This doctrine, still not entirely erased from sedimented science, is pre-phenomenological. Wherever located, shaped extension and colored shapes are equally *present* to microperceptual experience; but the epistemic organization of perception, as Michel Foucault's histories of perception have shown, is highly variable. Galileo himself makes this epistemic organization clear:

> As soon as I form a conception of a material or corporeal substance, I simultaneously feel the necessity of conceiving that it has boundaries of some shape or other; that relatively to others it is great or small; that it is in this or that place, in this or that time; that it is in motion or rest; that it touches, or does not touch, another body; that it is unique, rare, or common; nor can I by any act of imagination disjoin it from these qualities. But I do not find myself absolutely compelled to apprehend it as necessarily accompanied by such conditions as that it must be white or red, bitter or sweet, sonorous or silent, smelling sweetly or disagreeably; and if the senses had not pointed out these qualities, language and imagination alone would never have arrived at them. Therefore I think that these tastes, smell, colors, etc. with regard to the object in which they appear to reside are nothing more than mere names. They exist only in the sensitive body. . . . I do not believe that there exists anything in external bodies for exciting tastes, smells, and sounds, etc. except size, shape, quantity, and motion.[1]

Here we have the notorious doctrine of subject/object and the primary and secondary qualities. Today it is easy to deconstruct this archaic episteme, to see through it, as it were.

As Husserl saw, it is the isolation of one set of perceptual possibilities, those located in the geometrical metaphysics, that shapes the episteme of early modern science. It is now easy to see that Galileo plays loose with the interface between imagination and perception. To see is indeed to see within a field some set of related shapes, extensions, and bodies, but only when one turns to imaginative variations can such shapes be changed. In imagination, a square might as well be a circle or an amorphous blob of chaos. Nor can perception eliminate color; but again, when imagination enters, one can always presentify anything in different-from-actual colors. (I could argue, however, that there is no such thing as a colorless *visual* imagination, although the background field of visual imagination might be colored some indistinct color.) If this is so, the imaginative variations upon shape—any particular object can be indefinitely imaginatively reshaped, yet in whatever shape, there is a shape-presence within the visual presentification. But the same applies to color. And even if the color of the thing gets varied, to the extreme of a seemingly colorless or transparent object, there remains the same constancy of color-presence. The

---

1. Galileo Galilei, *On Motion*, trans. I. E. Drabkin (Madison: University of Wisconsin Press, 1960), p . 48.

background, however indistinctly colored (gray, shadowy, etc.), remains a constant color-presence. Microperceptually, at least, none of the secondary qualities are ever absent as qualities of the plenum. To isolate geometrical extension in a special abstractive or reductive act that is then given a special status is *preferred,* as Husserl recognized. Thus, Galileo had to take as his selection, by an act of abstraction, a favored set of qualities and dislocate the others to attain his geometrical "world."

Galileo substitutes, for part of the sensory plenum of the microperceptual lifeworld, his selection of an abstraction of the geometrical "world." This movement makes possible a new acquisition but simultaneously covers over an older one. If Husserl is right, how can one recover not what Galileo says of his vision but its secret relation to that seeing that opens the way to his new "world"? The answer, I would hold, lies in a sensitive historical reinterpretation of not what Galileo says about his theory but what must occur at the level of praxis. For this I turn from Husserl's to Derek Price's Galileo; in Price's Galileo one finds a science in close proximity to the optical continuum and its phenomenological structures.

Price returns the role of the telescope to Galileo's praxis. He notes that, independent of science as such, lens making had arrived at the point where strong diminishing lenses were possible. He holds that there is a strong and spontaneous link between philosophy of science and the history of technology. That link is the actual development of instrumentation:

> The archetype of this phenomenon is the earliest pathbreaking instance, namely, Galileo's use of the telescope. From the viewpoint of history of technology we can see that the telescope became possible when the crucial component—strong diminishing lenses—became commonly available during the late sixteenth century. Lens making had been an honorable and common craft for producing eyeglasses in the late middle ages; but lenses for myopic people were a late development, and very strong diminishing glasses had little commercial utility.[2]

Price's argument, a version on the historian's motto that the steam engine has more to do with the development of science than science with the steam engine, notes that the initial use of such strong lenses was in a domain other than science as such: "In the Renaissance, with its preoccupation with artist's perspective and other illusions, such perspective glasses attracted interest, since they showed the world in miniature like a microcosm."[3]

2. Derek de Solla Price, "Notes Towards and Philosophy of the Science/Technology Interaction," in *The Nature of Technological Knowledge: Are Models of Scientific Change Relevant?*, ed. Rachel Laudan (Dordrecht: D. Reidel Publishing Co., 1984), pp. 106–107.
3. Ibid., p. 107.

This echoes Michel Foucault's observation that the Renaissance was more concerned with spectacle than with science per se. But the very practice of creating perspectives within spectacles was to be taken into science. What must be noted is that the way of seeing which was spectacle seeing is part of a familiar paradigm for Galileo's time. But the same applies to the presence of lenses.

Lenses and lens use are a common, taken-for-granted phenomenon. They are part of Galileo's daily world, the macroperceptual world of the time. Moreover, there is a structural, epistemic relation between the Renaissance fascination with the microcosm and perspectives and the geometrical world Galileo helped make into a gestalt. Renaissance perspective, one of the many ways to represent spatiality, was already "geometrical." This is to say that a way of seeing is already part of the lifeworld that locates Galileo. It is not only a seeing but a praxis.

Price observes that a kind of fascination with lenses led to the compounding of strong diminishing lenses—the basis for the telescope—prior to Galileo's use of it. Indeed, the first uses were not at all designed for Galileo's purposes. The first telescopes were developed by Flemish lens makers, and there were both compound microscopes and compound telescopes by Galileo's day. Not unlike technology in our own day, however, the commercial hopes of the originators first were directed at the best commercial use: the military. Price notes, "As technologists, the lens makers tried to sell their invention as a military device to the richest Pentagon of their day. They were wrong, as inventors usually are, about the utility of their device. It was centuries before the telescope was of use in warfare—and then, only for signalling rather than for spying."[4]

Nor, according to Price, did Galileo deliberately first use his telescope for its eventual purpose. Again, in a striking glimpse of the early modern version of the science-industry-government combination:

> When Galileo duplicated the device, he had no idea that it was going to be of more interest to him than simply as a way of obtaining a commission from his Medici patrons. The Galilean telescope has such a tiny field of vision—like two keyholes in tandem a yard apart—that seeing anything through it is difficult.[5]

Yet he did look through it; and with his already selective vision, the macrovision of geometrical selectivity and perspective, Price argues:

> Most likely Galileo saw only the Moon on the first night and was perhaps not much more than amused to see the illusion that the Moon looked as if it had mountains on it instead of the face of a man in the Moon. The *geistesblitz* struck only when he looked again a few days later and saw

---

4. Ibid., p. 107.
5. Ibid., pp. 107–108.

that the shadows of the illusory mountains now looked completely different from the way they had earlier. Galileo knew enough astronomy to realize that a back-of-the-envelope calculation would enable him to compute the height of those mountains. He carried out the computation and found they were about the same size as mountains on the earth—this gave him what is sometimes called a "click"; everything fell into place. He knew the mountains were real and no illusion. He knew that for the first time in history he had seen something that could not be attested from common experience of the senses.[6]

In short, here we find, coming into proximity, Galileo's selective macrovision and his new microperception. The domain of shapes, measurements, angles gave Galileo the means to see the mountains of the moon as distantly similar to those of earth. But, contrary to his very depreciation of the senses, it was also the colored shadows (what if all "secondary qualities" had been invisible?) that made the measurements possible. The vision, albeit in the already secretly favored "black and white" of early modern geometrical thought, gestalted the plenum in a new way. Here is a dramatic paradigm shift within vision, but it is a paradigm shift that both is perceptual and involves *technology*. The technology of the telescope makes possible the perceptual shift itself.

Price argues that here lies the clue to the juncture between philosophy of science and the history of technology. Paradigm shifts are not simply intellectual revolutions. In my terminology, they are also new ways in which scientific vision can be embodied. Price extends his claim:

> The magnitude of this discovery cannot be overemphasized. That the Moon had mountains was an important discovery, but faded to relative triviality when compared with the nature of the experience itself. Galileo realized that he had manufactured for himself a revelatory knowledge of the universe that made his poor brain mightier than Plato or Aristotle and all the Church Fathers put together. This principle of *artificial revelation* was what was to worry the Church into behaving beyond the bounds of toleration and fair play toward a devout Catholic. . . . His little tube with lenses clothed the naked eye, allowing it to exceed all previous human experience.[7]

Galileo's exit from the Garden was dramatic. It took the shape of a new technologically embodied science in contrast to and far beyond the reaches of ancient Greek science. Devices for measurement, incidental to Greek science, were to become one of the central factors in a new set of perceptual relations to what was to become a greatly expanding world.

---

6. Ibid., p. 108.
7. Ibid., pp. 108–109.

What Galileo saw, he had *not* predicted nor sought; he saw a new astronomical world which included not only mountains on the moon but the phases of Venus, the satellites of Jupiter, and stars far in excess of the number available to naked vision. Price observes, again in a striking anticipation of the modern, publicity-prone science establishment, that Galileo immediately began a campaign to make his findings known. But most important, what Galileo's artificial revelation through optics accomplished was a *perceptual* way of establishing the new science simpler and more immediate than by calculation.

> Moving from the technological change to the scientific, let us consider the nature of Galileo's actual observations and their effect on theory. Clearly, since he did not know what to expect, he could not have begun his observations with the intent of testing Copernican theory. . . . [But, once seen,] these unexpected sights, of unquestioned [perceived] reality, made it obvious almost without argument and certainly without mathematics that the universe was Copernican.[8]

Price points out the well-known fact that on the basis of theory and efficient simplicity, Copernicus was actually no better able to account for stellar motion mathematically than Ptolemy; but the new technologically embodied observations helped settle the case. Thus, in addition to whatever Husserl might claim, Galileo also accomplished a new *technologically mediated* paradigm for scientific vision:

> Returning to Galileo, there is little doubt that the resounding success of the telescope . . . share(s) in the new way of doing philosophy by artificial revelations rather than by brainpower. . . . He has "looked on Nature bare" and learned from the experience that it could be done again and again. His insight was that the new aid to the senses was generalizable to other experiences; this changed thinking to the New Philosophy of experimental science. This was not the *testing* of theories but the trying out of new techniques to see what they would give, hoping for the unexpected.[9]

What Galileo says about his method of observation is belied by what he *does* in the observational situation. At the very least, the discovery of artificial revelation is as significant as the indirect mathematization of nature.

Although it is not as radical as Heidegger's claim, Price sees the tendency to emphasize the theoretical over the material conditions as a root of certain misconceptions concerning science itself. He argues that what we have called the nomological model of interpretation takes the function of experiment to be confirmation of theory. Here is a concrete example of how instrumentation plays a secondary and applied role within the older paradigm. Price locates this tendency to

---

8. Ibid., p. 109.
9. Ibid., p. 110.

downplay the role of instruments as a feature of the specific type of episteme associated with the old philosophy of science:

> I suggest that this unhappy position . . . has evolved from what now seems to me a misguided main line within philosophy of science in which it is supposed that the function of scientific experiment is in testing of hypotheses and theories. This supposition originated in an atypical period in the early nineteenth century when experiments in electricity and magnetism were common, and it continues to flourish because those who have their experience of science from books rather than from the laboratory bench naturally regard the history of science as if it were an intellectual enterprise, pure and simple.[10]

Price argues that it is much more the process of artificial revelation which lies in the linkage between modern science and technology which creates the conditions for subsequent paradigm shifts.

> Each radical innovation in this craft tradition gives rise not to the testing of new hypotheses and theories but rather to the provision of new information which affects what scientific theories must explain. This process, which I describe as 'artificial revelation,' is at the root of many paradigm shifts, perhaps not all, but most.[11]

In short, new instrumentation gives new perceptions, which, in a Kuhnian context, would minimally be the source of possible new anomalies. But more strongly, this new and now necessary link of instrumentation between science and technology is itself the extension of scientific vision which becomes a gestalt of an entire new context.

This focus upon the development of instrumentation, which Price takes as the connector between the philosophy of science and the history of technology, suggests another new interpretation of the role of technology in science. If Price is right, one can easily see that not only are Galileo's negative epistemic claims about the senses at best heuristic but that his own experience has a new gestalt in a new arrangement of the very primary *and* secondary qualities as mediated by technologies of observation. This leads Price to say:

> I claim a considerable advantage over Kuhn's account of paradigmatic revolutions and the philosopher of science's interpretation of experiment as a handmaiden for testing theories. I believe that the history of the craft of experimental science is the missing link between the history of technology and that of science.[12]

The same insight comes from a phenomenology of embodied vision.

---

10. Ibid., p. 105.
11. Ibid., p. 106.
12. Ibid., p. 110.

What the perceptual instrumentations do is place the observer in ever new positions with respect to the universe, whether at the macro or micro levels. The astonishment Galileo found in his early observations has been repeated many times in the history of astronomy. Only in this century have many new both anomalous and varied gestalt phenomena been "seen." Moreover, much of modern astronomy owes its deeper penetration to a whole added multiple-sensory dimension. Radiotelescopy, an instrumental "artificial" ear, began several decades ago to "hear" some kind of stellar activity in previously "dark" or "empty" areas of the heavens. This artificial revelation is the high-technology version of the much more ordinary experience of hearing some rustle in the dark, which only later is visually identified (maybe with the aid of a flashlight). It is not only new theory that separates us from any of the ancients but also the newly extended perception that occurs *through* instrumental enbodiment. To cement the point inversely, take all instrumentation away from the scientific community and then ask what it would and could know. Its limits would very quickly reduce to precisely those admirable but at best speculative notions of our Greek forefathers.

Galileo's discovery, then, was not only the indirect mathematization of nature but also, in Price's terms, *the artificially aided perception of nature*. His perception, now incommensurate with that of any of the ancients or the Church Fathers, stood at the forefront of the tradition of modern *technologically embodied* science that characterizes our own time.

If the focus that lies at the center of Galileo's new science is perception embodied by artificial revelation, the field within which it occurred is that of a lifeworld.

The historians—Price, Lynn White, Jr., and earlier, Lewis Mumford—have made us much more keenly aware of the rapid accumulation of "modern" technologies that first the Renaissance and then the early modern scientists would take for granted. But this awareness may also be adumbrated by a closer phenomenological look at what lies latent in praxis, in cultural macroperception.

First, note something about the texture of seventeenth century technological life. The sheer extent of inventions is impressive.

Optics, which led to the astronomical and later to the microcosmic perceptions of previously unsuspected aspects of the universe, began as early as the eleventh century. Optical technology was common in spectacles by the thirteenth, and lenses were compounded for telescopes and microscopes by the seventeenth.

There were machine works, more often in wood than in iron for large works but in iron for smaller works, which well predated the rise of modern science. The lowlands were pumped; the cathedrals built; there were large cranes, elevators, wind and water mills—all common from the thirteenth century on. Nor should one forget the often primary user of technologies, the military. Cannons appear in the twelfth century and were sophisticated by Galileo's time; and rifled firearms were in use. There was even the metered taxi. In short, the technological texture of postmedieval Europe was impressive. It was a part of the daily life of at least every urban dweller.

Furthermore, it had been enriched by exploration and commerce, which began prior to Columbus's discovery of the New World. Gold and spices had been flowing into Europe for several centuries before Galileo saw the chandelier as pendulum. Such was the relatively technologically sophisticated lifeworld of the late Renaissance. A historically sensitive look at the time of Galileo shows the impressive array of already common technologies, but it also suggests that this available infrastructure had acquired the shape of a common form of life.

The array, however, is too complex to account immediately for the praxis that becomes concentrated in the birth of science. I must turn to a narrower field of analysis. I have already suggested some of the ways in which optical technology made a new vision possible. What Price calls artificial revelation and I called *embodied vision* took shape in many quarters. Renaissance *visualism*, I suggest, although it historically precedes Galileo's use of it, was already well established as the privileged sense; it was transformed from the theatrical and artistic perspectives of the Renaissance into early modern science. The "laws of perspective" were already geometrical interpretations of depth. These were known by 1440. And Leonardo da Vinci, who was much more engineer than scientific theorist, had already transformed an older tactile anatomy into his artistic and Michelangelistic *visual* anatomy. The fascination not only with the microcosm but with bird's-eye positions of observation were virtually *de rigueur* as the perspective of the fifteen and sixteenth centuries. In short, the reduction of microperception to a favored position for vision had already begun. Moreover, the way artists looked through geometrical grids had laid the groundwork for that modification of earlier vision. Galileo could take a lot for granted in the already established shaping of perception.

To gain a closer view, however, I shall examine a few special technologies, particularly as they relate to the senses and to the perception of space and the perception of time. I begin with time perception and clock technologies and their role in the lifeworld.

A. LIFEWORLD TECHNICS: TIME PERCEPTION

St. Augustine's famous landmark in the philosophy of time preceded the era of clock culture: "What, then, is time? If nobody asks me, I know; but if I try to explain it to one who asks me, I do not know." This statement seems to imply something like a direct, if non-expressible, intuitive access to time. Yet the lifeworld is filled with perceptual time indicators, which in modern cultures are dominantly clocks.

The clock has long been recognized as an important technology in the West. Lewis Mumford's chapter in *Technics and Civilization* on "The Monastery and the Clock" (1934) is probably one of the best-known expositions of a relationship between a technology and a form of life. He claims that

> The clock, not the steam-engine, is the key machine of the modern industrial age. For every phase of its development, the clock is both the outstanding fact and the typical symbol of the machine: even today no other machine is so ubiquitous. Here, at the very beginning of modern technics, appeared prophetically the accurate automatic machine which, only after centuries of further effort, was also to prove the final consummation of this technics in every department of industrial activity.[13]

Mumford's exposition primarily related the clock to a changed social time. First, in the monastery, the clock regulates social movement, making it both more ordered and more uniform. The first clocks used in the monasteries were one-handed and told only the *hours*. These, in turn, regulated the hours of work and devotion of the monks. Mumford notes that this also changes our perception of time, at least in contrast with non-clock cultures. One such change is increasing *quantification*. "The application of quantitative methods of thought to the study of nature had its first manifestation in the regular measurement of time; and the new mechanical conception of time arose in part out of the routine of the monastery."[14]

The same point was made even earlier by Martin Heidegger in *Being and Time* (1927). Like his tools, technologies "take account" of nature in some way. The covered railway platform takes account of bad weather; lighting, of the dark; and the clock, of the rhythms of time. Time, Heidegger argues, is foundationally existential. It is based upon the human sense of finitude, and time accounting is the accounting for what one can do with the time of finitude; hence, temporality arises out of *concern*. This, he argues, underlies both clock and non-clock cultures, but in both cases time is figured in relation to what humans concretely encounter in the world. In non-clock cultures, these are primarily phenomena such as the rising and setting of the sun and the movement of the seasons. There is an almost natural analogue between natural movement and the clock. "And because the temporality of that Dasein which must take its time is finite, its days are already numbered."[15]

We have already noted how Heidegger takes the realm of equipment as that which is closest to ordinary life. The same applies to

13. Lewis Mumford, *Technics and Civilization* (New York: Harcourt, Brace and Co., 1962), p. 14.

14. Ibid., p. 12.

15. Martin Heidegger, *Being and Time*, p. 466.

clocks. For early Heidegger, clocks are based upon existential temporality. "Temporality is the reason for the clock. As the condition for the
possibility that the clock is factically necessary, temporality is likewise
the condition for its discoverability."[16] Heidegger's time, parallel to
Mumford's, is also social time. Time is intentional insofar as it is always
directed; it is time for _____ , time-in-order-to _____ , time for
which, etc. "Only now, in any case, can the time with which we concern ourselves be completely characterized by its structure; it is dat-
able, spanned, and public; and as having this structure, it belongs to
the world itself."[17] In this sense, clocks reflexively point back to a
primitive existential base.

    Clocks, however, once developed, transform the perception of
time. For example, measurement itself becomes a possible focal activity. Clocks are the condition for making such measurements more explicit, even if founded upon existential temporality. But with the
emergence of measurement which takes on its own trajectory of possibility, clocks may also within social embeddedness open the way to a
separation from the social time of non-clock culture:

> Comparison shows that, for the "advanced" Dasein, the day and the
> presence of sunlight no longer have such a special function as they have
> for the "primitive" Dasein on which our analysis of "natural" time-
> reckoning has been based; for the "advanced" Dasein has the
> "advantage" of even being able to turn night into day. Similarly, we no
> longer need to glance explicitly and immediately at the sun and its
> position to ascertain the time. The manufacture and use of measuring-
> equipment of one's own permits one to read off the time directly by a
> clock produced especially for this purpose.[18]

✓ Here is a temporal, technological exit from the Garden.

    The clock not only changes what is emphasized in time account-
ing, but as an acquisition of the lifeworld, it makes possible other
moves that differ from non-clock time accounting. Heidegger notes
that nature is read differently via the clock: "Our understanding of the
natural clock develops with the advancing discovery of *Nature* and in-
structs us as to new possibilities for a kind of time-measurement which
is relatively independent of the day and of any explicit observation of
the sky."[19] In Heideggerian terms, the clock both changes a variable in
time accounting (it makes measurement itself a theme of temporal dimensionality) and opens the way to a type of autonomy that partially
distances from "nature." Both of these acquisitions Galileo could take
for granted. They already were part of the form of life characteristic of

---

16. Ibid., p. 467.
17. Ibid., p. 468.
18. Ibid., p. 468.
19. Ibid., p. 468.

the seventeenth-century lifeworld. Lynn White, Jr., has also noted how central the clock has been to the Western, even more specifically the Latin Western, awareness. He notes that clocks were very differently embedded in cultural praxes in the Latin West compared to the Byzantine East.

> In a separate building outside Hagia Sopia, Justinian placed a clepsydra and sundials, but clocks were never permitted within or on Eastern churches; to place them there would have contaminated eternity with time. As soon, however, as the mechanical clock was invented in the West, it quickly spread not only to the towers of Latin churches but also to their interiors.[20]

White lays this difference to the already Baconian desire for power in the medieval West. This thirst, evident, he claims, in the voyages of discovery and conquests of the fifteenth and sixteenth centuries and actualized in the technological conquest of non-human power through machinery, has a long history. But White also recognizes an important movement of lifeworld acquisition.

Once time could be perceived *through* clock perception, both measured and distanced from natural "clocks" such as the sun or seasonal or heavenly bodies in motion, one could also take the mechanical motion as *paradigmatic*. Indeed, to understand heavenly motion as a "clock" is anachronistic, but it is or may be a metaphor for focusing understanding itself. And this movement occurred early, long prior to Galilean science:

> It is in the words of the great ecclesiastic and mathematician Nicholas Oresmus, who died in 1382 as Bishop of Lisieux, that we first find the metaphor of the universe as a vast mechanical clock created and set running by God so that "all the wheels move as harmoniously as possible." It was a notion with a future: eventually the metaphor became a metaphysics.[21]

Nature herself was thus read as though a clock, her ontology so interpreted, even if the clock arrived late upon the scene. Nicholas was very quick to adapt this metaphor-metaphysic, since clocks with divisions into minutes and the perfection of the mechanical clock have been dated 1345 and 1370, respectively. One might suspect that this quick fascination with a human artifact hides a temptation that still persists in the late twentieth century. In our time, not only the universe but also ourselves are read through our technologies.

These preliminary observations are indicative of the pervasive role

---

20. Lynn White, Jr., "Cultural Climates and Technological Advance in the Middle Ages," *Viator* 2 (1971), p. 171.

21. Ibid., p. 125.

of the clock within the lifeworld of early science. The implicit phenomenologies of the history and philosophy of clocks can be made
more explicit. First, reading time through a clock substitutes *what* is
read; one reads the clock rather than the heavens. Even if the rhythms
of nature are not totally abandoned, they increasingly take on a background rather than foreground significance. Second, the type of motion of the clock itself (until recent electric clocks) takes on a different
form, one noted from the beginning; early clock motion was "jerky"—
it ran by mechanical jumps. Turing's description of a discrete state machine, the very model for a computer, applies quite well to the medieval clock:

> [Discrete state machines] move by sudden jumps or clicks from one quite
> definite state to another. These states are sufficiently different for the
> possibility of confusion between them to be ignored. Strictly speaking,
> there are no such machines. Everything really moves continuously. But
> there are many kinds of machines which can profitably be thought of as
> being discrete state machines.[22]

Medieval clocks progressed by hands that jumped from one instant to
another, and the same jumping motion occurs in today's quartz versions.

Oresmus in the 1300s likened the heavens to a clock; another
metaphorical use was practiced by Galileo. He is reported to have
used for measuring time in his acceleration experiments not a clock
but his pulse. But he had already likened his pulse to the clock; that is,
he read his pulse as a clock, thus indicating that the digital or quantitative had already become his way of reading time.

Third, the precision and regularity of the clock, in spite of the analogues in the metaphor of the harmonious running of the spheres, also
contrasts with the heavenly motions. In fact, once the regularity of a
clock became instrumental both at the micro- and macroperceptional
levels of early modern scientific measurement, the very discovery of
heavenly anomalies became possible. The "perfect" circular and regular clocklike motion was soon discovered not to be the motion of the
heavens. Kepler's discovery of elliptical motions and the later discoveries of perturbations in planetary motions are anomalies contrasting
with the instrumental motion suggested or actually instantiated by
clocks. One learns, in this sense, as much by the difference between
the instrumental "intentionality" as by the similarities which usually
first dominate lifeworld acquisitions.

Both circular and discrete representations of clock reading would
separate the New Adam from any monodimensional perception of the

---

22. A. M. Turing, "Computing Machines and Intelligence," *Minds and Machines*, ed.
Alan Ross Anderson (Englewood Cliffs: Prentice Hall, 1964), p. 11.

Garden; but the separation is neither total nor without the possibility of one kind of recovery. One can nakedly view the motions of the sun and the stars, as many have done in the twentieth century for special purposes, but once one has left the Garden the past ways of gaining knowledge are more often lost or forgotten. Very few of us "know" what any peasant would have known about planting times or take seriously in the same way the passages of the seasons. Our praxes have irreversibly changed.

Note, now, several aspects of time perception in clock reading: The perception of time through a clock is a *hermeneutic* perception. One "reads" the clock and, *indirectly* through it, time; but as with all reading, the "intentionality" or the inherent pregiven "interpretation" of time is already selective. First, note that the representation of time on the clock face is spatial-perceptual as well as temporal. Were it possible to simply reduce temporal movement to one feature in perceiving clocks, the already noted odd digital or jerking time motion might well be selected. But time is always space-time, so in a more fundamental sense the quasi-spatiality of time representation is not itself anything like a total distortion. When this is taken into account, one may perceive that traditional clocks represent *two* dimensions of time *simultaneously* and *visually*.

One dimension is the *instant* of time. This is the representation located analogically in the precise position of the hands. In one-handed clocks, that instant itself could be considered somewhat "broad" in that the hand could be positioned in some click, upon or between the hours. The other dimension is the *span* or duration of time, which is represented by the clock face, whether circular (as in some very ancient water clocks) or as a partial circle or arc (as in sundials).

Moreover, the instant of time—where the pointer rests—is located within or upon the field of durational representation of time. These two dimensions remained constant in clock evolution until recently. But there have been refinements in the evolution of clocks that point up a particular *trajectory*. That trajectory follows precisely those tendencies already noted by both Mumford and Heidegger.

One-handed clocks, which discriminated between hours, were superseded quite early by two-handed clocks, which further divided or quantified time into minutes. Two things occur in this refinement of measurement: First, not only is time more finely quantified into smaller units, but in a very subtle sense, one can say that minutes gradually become more "important" than the larger hour unit. The trajectory is *towards* the instant, making it more focal and consequently placing the duration of time more in the background.

If one wants to know "exactly" what time it is, one looks at the minute hand—but still simultaneously at the hour hand and field. One can *see* at a glance, in the familiar visual gestalt, that it is ten after twelve. Later still, a third hand was added, and an even further precising of the instant became possible—the *second*. I need not add the importance of this gradual precising of the instant for the praxis of science and the art of measurement, other than to indicate that this is yet another exemplification of *technological*-science as it emerges historically.

The trajectory toward the primacy of the instant could also be illustrated in terms of public time. The time analyses of the Industrial Revolution, in which time-motion studies became the very exemplars of measure, and atomized or digital time in factory praxis make the point concerning time perception and use. One could note an even more ordinary experience. How long is the anxiety quotient if one compares it in clock and non-clock cultural space? Take a simple luncheon appointment: The non-clock inhabitant may agree to meet the other "under the Baobab tree when the sun is high in the sky." The first to arrive will not experience waiting as anxiety or fretting for a long duration of time, up to hours in some cases, because the time sense here is not only not focused upon the instant but because the entire cultural time gestalt is looser. Contrarily, the businessman in the Trade Tower will likely become anxious within a matter of minutes, certainly in much less than a half hour, if his client is late.

The terminus of this trajectory is reached with the introduction of the common digital watch. Now the instant, and with it the jerk from one number to the next, was so enhanced and emphasized that the representation of the time field or duration disappeared entirely. It was soon recognized that such a device was not only a *magnification* of one dimension of time but the virtual elimination and displacement or total perceptual *reduction* in the representation of time. In rejecting this loss, people chose either analogue or both digital and analogue models again. In passing, one may also note that the reading process itself changes. An analogue watch fits the visual gestalt perception—it is read in a glance in which both instant and field are simultaneously given. The digital watch represents only the instant, and thus the duration must be *inferred,* thereby calling for a different mental process.

Clock cultures read time through the clock. Time perception is a mediated, hermeneutic perception. We have noted that for the European West this time praxis was already sedimented prior to the Renaissance and the rise of modern science. And it is a major factor in cultural differences, particularly those that contrast with sun-clock cultures.

B. LIFEWORLD TECHNICS: SPACE PERCEPTION

If Galilean science could presuppose the context of clock culture in its subtle transformations regarding time, so could it assume a technologically mediated and mathematically interpreted perception of space. Once again I am suggesting that Galileo was, in fact, late drawing certain consequences to an already established lifeworld praxis—although this in no way casts doubt upon his scientific genius.

The example of space perception here is found in long-distance navigation. By world standards, Europeans were rather late trans-oceanic navigators. South Pacific navigators actually spread over the Pacific a millennium before even Leif Ericson; the Chinese were the trade masters of the northern Pacific and the Indian oceans; and the Arabs covered the oceans centuries before Christopher Columbus found the New World. Also preceding the voyages that were to set our own histories, there are evidences of accidental long voyages such as those that stranded Eskimos in Ireland in medieval times and left the Roman galleon recently found off the Brazilian coast. It is also likely that many prehistoric migrations occurred by boat. European transoceanic voyaging came relatively late.

Stimulated by the hunger for goods that had fed the appetites of Europeans since the Crusades, early navigators practiced coastal navigation for centuries before Columbus. But coastal navigation is largely an "eyeball" operation. Ships need not go far off shore; they may anchor or harbor in foul weather; and they may make a leisurely and thus relatively safe voyage. This is not to say that out-of-sight techniques for certain passages were lacking; they may be traced back to Phoenician, Greek, and Roman times. But long-term transoceanic navigation poses much more difficult problems of spatial orientation.

Apart from legends, the first documentable such voyage was that of Leif Ericson and his kin, the Vikings who discovered Greenland and Newfoundland and there developed colonies. Unfortunately, little is known of the navigational methods used, but there are some hints that as early as these voyages a primitive form of technological mediation may have been employed.

Polaris, the North Star, was, of course, familiar. For Northern Hemisphere sailors, this fixed star provides a constant not available to the Southern Hemisphere. A second fortuitous phenomenon is the magnetic north pole, which roughly corresponds to the heavenly position of Polaris. Such fixed points can be very useful for space orientation and were so used in ancient times. There are hints that the Vikings may have had a primitive lodestone. It would have been a technological-mediating device. (The Chinese definitely had the compass and made it central to much of their navigational technique.) Such a device, like the clock, allows a certain distancing from immediate natural phenomena. For example, in a fog when neither the sun nor the North Star could be seen, the lodestone would still provide a constant by which to steer.

It is remotely possible that the Vikings had a device previously unknown to us, the *bluestone*. Viking ships have frequently been uncovered, and within the clutter found inside the hulls were various stones. Large ones would of course be ballast. But amidst these large stones were a number of flat, disk-like stones, striated with parallel lines of quartz or some other mineral. It has been suggested that these stones were used in haze or light fog to locate the sun—apparently they refract visible effects, even when the sun can't be seen. An acquaintance of mine, an SAS pilot, claims to have used these bluestones in just such a way and confirms that they do in fact work. Such aids to navigation would have meant that the Vikings were already on what was to become the dominant European trajectory of technologically mediated navigation.

If we take Columbus as our example, however, that conclusion is clear, and it parallels the transformation of time perception with the clock—European transoceanic navigation was from then on both mathematical and technologically mediated. He used charts divided into the world grids similar to those used today. From the thirteenth century on, Portulans, which are charts of the coasts divided by dominant wind patterns, were familiar. Here the bird's-eye perspective was already established, a practice which contains subtle but deep implications for spatial orientation. Second, Columbus calculated his latitude through astronomical sightings and gauged his positions accordingly (longitude measurement had to await the centuries until a seagoing clock accurate enough to mediate the measurement). This calculation was aided by instruments—at the least, a crude astrolabe. Other instruments of measurement were also common, such as the log, a rope in which knots were tied and at the end of which was some type of log. When thrown overboard, the speed of the ship could be calculated by how long it took for the knots to go by. "Log" and "knot" became naval terms from this device.

In short, Columbus employed, in familiar praxis, a series of instruments to calculate in crudely mathematical terms where he was. The peculiarity of this praxis will not become apparent until variations can establish quite contrary modes of navigation; but in the immediate context, it is enough to note that a technologically mediated form of space perception was already familiar a century before Galileo and was therefore part of the daily lifeworld at least of the navigators of the time.

The tensions over how far the Orient must be arose on board not so much because individuals feared falling off the edge of the world (all educated people knew the earth was round) but because the size of the earth was seriously underestimated. This belief in smallness was not to be corrected for some centuries, but the belief may be related to the previous underestimations of the heavens by those prior to the Greeks. I suspect that the foreshortening which belongs to microperception plays some role in this belief phenomenon.

If Columbus was already "technological" in his navigational science, it remains to detect some of the subtle lifeworld aspects of the perceptions which belong to this praxis. There is a very peculiar aspect which may be found in European (and later all Western) navigational praxis regarding the *from which,* or position from which, location is determined. I refer to what may be called the bird's-eye position. Columbus did not declare himself to be simply "here" (on board, incarnate, located within the immediate perceptual surrounding). Rather, he "read" himself to be at some position upon a chart—as if he were somehow above the ship he was traveling in and judging its position from a *disembodied* or imaginative perspective that he did not in fact occupy.

I have stated the case deliberately in this fashion, because the assumption of a bird's-eye perspective is so familiar to us that we automatically follow its peculiarities as a praxical second nature. The situation can be more carefully analyzed if we attend to a phenomenology of chart reading.

A chart or map is read literally *from above.* Insofar as it represents the terrain, that terrain is seen *from the heavens.* It is to "see" the earth *from a position I do not actually occupy.* And because this is so, to make chart reading intuitive—to constitute the intuition—I must learn to make a *hermeneutic shift.* Imagine our Columbus with a chart of unknown waters (see Figure 3).

He knows, from previous experience and from the cartography of the day, the coastal features of Europe. He sails westward along a roughly parallel latitude, charting his progress with distance estimates. But he continues to locate himself on the chart, *from above.* Now, this is his actual *reading position* and to that extent the microperception is embodied—but it is *not* his actual position with respect to the immediate environment of the ocean. To establish that, he must correlate the bird's-eye position with a second, embodied position; and this calls for a specific *hermeneutic act.*

The point can be made clearer through a series of possible map variations. Maps need not be constructed from bird's-eye perspectives. Some maps are simply reading directions: "To reach my house from the university, take the back exit out of the university, turn left onto Nicolls Road until it reaches the stop light, then turn right onto 25A." Here, a written instruction functions as a map. Or a map could be drawn to represent, still-photo-like, a series of perceptual representations which would be isomorphic with respect to the driver's position and the surrounding countryside. In this case, I could have instructed the driver to look in sequence at a set of photos I give him showing first—from the driver's position—the exit to the university, then the stoplight at the intersection, etc. Such a series would isomorphically represent what would be seen *from* the driver's seat at such-and-such a stage of the journey.

Our usual maps or charts, however, do something else. They establish position by requiring a hermeneutic perception, and the chart position must be "read" onto our bodily, perceptual position. The difficulty this entails is obvious to anyone who has sailed far off shore, particularly in obscure conditions and in unfamiliar territory. What the chart reveals is depicted from above, and now one must translate— that configuration there must be this tower here (?). Were one to read the chart from the position of an airplane at low altitude, there would be more of a template isomorphism. Yet we do such hermeneutic acts in quite ordinary contexts. The very reading process has become an acquisition of our lifeworld and makes habitual one aspect of our very notion of perspective.

In Columbus's case, this brief description should be sufficient to show that location was already geometrically established; that the establishment of such locations entail hermeneutic perceptions; and that within the hermeneutic context, instrumentation is used to mediate the perceptions. This is a practical parallel to the techniques of the Renaissance regarding perspective and geometrization. Columbus and da Vinci were, after all, contemporaries. This praxis was already common a century before Galileo utilized it in a new context. It was part of the lifeworld that he could assume and take for granted.

C. ARTIFACTS AND TECHNOFACTS

Our actions are embedded in the multiple ways we interact with and presuppose our technologies, yet this multiplicity remains perceptually and praxically ambiguous. What is there about the artifacts we employ? The human-technology juncture displays a puzzling ambiguity. Imagine there to be, lying on my coffee table, a symmetrically shaped stone. It is ovaloid in form with one end more pointed than the other, flattened by chipping and flaking such that a sharp edge runs around the entire circumference. My guests arrive; and after the first round of drinks, someone picks up the stone and asks, "What is it?"

The artist at the party takes it and pronounces that it is an *objet trouvé*. It is a kind of art object. A writer, noticing that the breeze is now ruffling the pages of the magazine on which the stone had been placed, counters that the stone is merely a paperweight, a practical object. But the anthropologist scoffs, and with Sherlockean deductions delivers the assertion that it is, *in reality,* an Acheulean hand axe. He notes that not only is it shaped right but that objects of this sort are familiar parts of the Stone Age tool kit, found from the banks of the St. Acheul River in France to the Vaal River in South Africa.

The difficulty under this party game is genuine. What *is* the object? The object is or could be any of the things named, or it could become how it is used. That, too, is an ambiguity essential to technologies. As object, the Acheulean hand axe could belong to any number of use contexts; but, the anthropologist retorts, it was *designed* to be a hand axe, and it has been so taken by decades of classifiers in the field.

Then, however, another informed guest mentions having read a piece from a recent *Natural History Magazine* about this very axe. It seems, after decades of accepted notions about its designed use, that there was an anomaly for a hand axe—the circumference sharpness would surely be dangerous to the user. In a mini-gestalt shift, the author proposed that perhaps the "axe" was actually a projectile, to be thrown like a discus. And in good contemporary scientific practice, the author asked discus throwers to throw the object and, behold, the sharp, pointed end landed down in 31 out of 45 throws! A new interpretation of designed use comes into being.

What we have in this example would be recognized by the literary humanist as the anthro-archeological counterpart of the *intentional fallacy*. In literary circles, it has long since ceased to be fashionable to seek an author's "intentions" in writing a story; indeed, it has been argued that it is not even possible to do so. In this case, the designer's intentions are sought but remain under question. Similarly, in any good interpretive exercise (hermeneutics), the humanist would recognize that while there were intentions, the story belongs now to a different context and possibly to multiple contexts. So, too, does the stone. The designer's intentions play only a small part of the subsequent history of the artifact. It was, after all, Nobel's intention in the invention of dynamite that it be used for mining and the benefit of humankind. Design, in the history of technology, usually falls into the background of a multiplicity of *uses*, few of which were intended at the outset.

At an even deeper level, this multiplicity of uses reveals a beginning phenomenological clue that must be followed. There is no "thing-in-itself." There are only things in contexts, and contexts are multiple. One could easily go back to our ex–hand axe and beyond the now-imagined primal story of Stone Age hunters hiding in the rushes of ancient rivers, emerging as a group "artillery" shelling or stoning a herd of antelope or a flock of geese, now placing the stone in any number of contexts. I could have used it as a keystone in one of my Vermont fireplaces; my Viking could have placed it into the hold of his vessel as ballast; or one could have cemented it, point up, along the top of a stone fence, as the Europeans do.

The ambiguity of the Acheulean object is the ambiguity of technology in general. This is another reason for not beginning with the
object as object. A technological object, whatever else it is, *becomes*
what it "is" through its uses. This is not to say that the *technical* properties of objects are irrelevant, but it is to say that such properties in
use become part of the human-technology relativity. Nor is it to deny
that there is a specific type of history to the development of technical
properties.

Animal technics are frequently temporary use of simply found objects (thorns, sticks, etc.), and this was doubtless the case with early
humans. But sticks do not remain sticks. They become spears, and in
the process there is a shaping and manufacturing into technological artifacts. Australian Aboriginals regularly plucked the long, straight stems
of certain plants for their spears, but they then fastened fire-hardened
hardwood tips to the ends of the shafts with a vegetable glue and vegetable bindings. They began to transform the simple "found" quality
of such objects.

Higher order or more complex transformations have also occurred
in the past but have been accelerated in today's high-technology contexts. What I shall call a *technofact* is an object in which the very materials themselves have undergone levels of transformation. Plastics,
now pervasive, simply do not occur in nature. (Although similar sub-
stances can approximate plastics. I remember seeing the black and
shiny floors of traditional Zulu beehive huts and noting how much like
a plastic these hard materials were. When I asked how they were
made, the answer was "ant heap and cow dung.")

Once the technological trajectory has begun, artifacts and
technofacts become increasingly differentiable from natural products.
But this too is a sign of the entry into the human-technology relativity.
Even the technical properties take on significance in the use context.
Stone, once shaped, enters praxis. Hardness becomes hardness for
_____ ; shapability, shapability for _____ . Thus what is "natural" in
the stone becomes artifactual within the relativity.

If the ambiguity of the object is one side of the problem, then the
other side is that virtually any object may become a technology—at
least, if it can be brought into the range of human praxis. And if our
hand axe could have been or could become a use object in a range of
use contexts, the inverse is also the case. Any use context can also utilize any number of artifacts or technofacts. If Heidegger's hammer
could be used as a hammer (its designed purpose) but also as a mur-
der weapon, an objet d'art, a pendulum weight, etc., so could any
number of items be used to hammer, including the Acheulean
hand axe.

This ambiguity of objects in use contexts is abundantly illustrated
in the histories of *technology transfers*. Lynn White's Hindu prayer
wheel (a windmill which is an automated prayer machine—the prayer

goes to the heavens every time the wind turns the wheel) becomes a source of power to pump the Dutch lowlands. The Spanish conquistadors quite consciously took hawksbells (small brass bells used in falconry) and mirrors, both designed for different purposes, as baubles for trade and fascination to the Arawaks and Caribs of Central America and the Caribbean. When the object is transferred, its use may be very different from that of its previous cultural context.

All of these ambiguities may be noted prior to a more rigorous phenomenology of human-technology relations. The purpose of such a phenomenology, however, is to reduce the open ambiguity noted to a structural analysis which cuts across and accounts for the range of possibilities that occur within the essential ambiguity of technology.

5. Program One: A Phenomenology of Technics

The task of a phenomenology of human-technology relations is to discover the various structural features of those ambiguous relations. In taking up this task, I shall begin with a focus upon experientially recognizable features that are centered upon the ways we are bodily engaged with technologies. The beginning will be within the various ways in which I-as-body interact with my environment by means of technologies.

A. TECHNICS EMBODIED

If much of early modern science gained its new vision of the world through optical technologies, the process of embodiment itself is both much older and more pervasive. To embody one's praxis *through* technologies is ultimately an *existential* relation with the world. It is something humans have always—since they left the naked perceptions of the Garden—done.

I have previously and in a more suggestive fashion already noted some features of the visual embodiment of optical technologies. Vision is technologically transformed through such optics. But while the fact *that* optics transform vision may be clear, the variants and invariants of such a transformation are not yet precise. That becomes the task for a more rigorous and structural phenomenology of embodiment. I shall begin by drawing from some of the previous features mentioned in the preliminary phenomenology of visual technics.

Within the framework of phenomenological relativity, visual technics first may be located within the intentionality of seeing.

I see—through the optical artifact—the world

This seeing is, in however small a degree, at least minimally distinct from a direct or naked seeing.

I see—the world

I call this first set of existential technological relations with the world *embodiment relations*, because in this use context I take the technologies *into* my experiencing in a particular way by way of perceiving *through* such technologies and through the reflexive transformation of my perceptual and body sense.

In Galileo's use of the telescope, he embodies his seeing through the telescope thusly:

Galileo—telescope—Moon

Equivalently, the wearer of eyeglasses embodies eyeglass technology:

I—glasses—world

The technology is actually *between* the seer and the seen, in a *position of mediation.* But the referent of the seeing, that towards which sight is directed, is "on the other side" of the optics. One sees *through* the optics. This, however, is not enough to specify this relation as an embodiment one. This is because one first has to determine *where* and *how*, along what will be described as a continuum of relations, the technology is experienced.

There is an initial sense in which this positioning is doubly ambiguous. First, the technology must be *technically* capable of being seen through; it must be transparent. I shall use the term *technical* to refer to the physical characteristics of the technology. Such characteristics may be designed or they may be discovered. Here the disciplines that deal with such characteristics are informative, although indirectly so for the philosophical analysis per se. If the glass is not transparent enough, seeing-through is not possible. If it is transparent enough, approximating whatever "pure" transparency could be empirically attainable, then it becomes possible to embody the technology. This is a material condition for embodiment.

Embodying as an activity, too, has an initial ambiguity. It must be learned or, in phenomenological terms, constituted. If the technology is good, this is usually easy. The very first time I put on my glasses, I see the now-corrected world. The adjustments I have to make are not usually focal irritations but fringe ones (such as the adjustment to backglare and the slight changes in spatial motility). But once learned, the embodiment relation can be more precisely described as one in which the technology becomes maximally "transparent." It is, as it were, taken into my own perceptual-bodily self experience thus:

(I-glasses)-world

My glasses become part of the way I ordinarily experience my surroundings; they "withdraw" and are barely noticed, if at all. I have then actively embodied the technics of vision. Technics is the symbiosis of artifact and user within a human action.

Embodiment relations, however, are not at all restricted to visual relations. They may occur for any sensory or microperceptual dimension. A hearing aid does this for hearing, and the blind man's cane for tactile motility. Note that in these corrective technologies *the same structural features of embodiment* obtain as with the visual example. Once learned, cane and hearing aid "withdraw" (if the technology is good—and here we have an experiential clue for the perfecting of technologies). I hear the world through the hearing aid and feel (and hear) it through the cane. The juncture (I-artifact)-world is through the technology and brought close by it.

Such relations *through* technologies are not limited to either simple or complex technologies. Glasses, insofar as they are engineered systems, are much simpler than hearing aids. More complex than either of these monosensory devices are those that entail whole-body motility. One such common technology is automobile driving. Although driving an automobile encompasses more than embodiment relations, its pleasurability is frequently that associated with embodiment relations.

One experiences the road and surroundings *through* driving the car, and motion is the focal activity. In a finely engineered sports car, for example, one has a more precise feeling of the road and of the traction upon it than in the older, softer-riding, large cars of the fifties. One embodies the car, too, in such activities as parallel parking: when well embodied, one feels rather than sees the distance between car and curb—one's bodily sense is "extended" to the parameters of the driver-car "body." And although these embodiment relations entail larger, more complex artifacts and entail a somewhat longer, more complex learning process, the bodily tacit knowledge that is acquired is perceptual-bodily.

Here is a first clue to the polymorphous sense of bodily extension. The experience of one's "body image" is not fixed but malleably extendable and/or reducible in terms of the material or technological mediations that may be embodied. I shall restrict the term embodiment, however, to those types of mediation that can be so experienced. The same dynamic polymorphousness can also be located in non-mediational or direct experience. Persons trained in the martial arts, such as karate, learn to feel the vectors and trajectories of the opponent's moves within the space of the combat. The near space around one's material body is charged.

Embodiment relations are a particular kind of use-context. They are technologically relative in a double sense. First, the technology must "fit" the use. Indeed, within the realm of embodiment relations one can develop a quite specific set of qualities for design relating to attaining the requisite technological "withdrawal." For example, in handling highly radioactive materials at a distance, the mechanical arms and hands which are designed to pick up and pour glass tubes inside the shielded enclosure have to "feed back" a delicate sense of touch to the operator. The closer to invisibility, transparency, and the extension of one's own bodily sense this technology allows, the better. Note that the design perfection is not one related to the machine alone but to the combination of machine and human. The machine is perfected along a bodily vector, molded to the perceptions and actions of humans.

And when such developments are most successful, there may

arise a certain romanticizing of technology. In much anti-technological literature there are nostalgic calls for returns to simple tool technologies. In part, this may be because long-developed tools are excellent examples of bodily expressivity. They are both direct in actional terms and immediately experienced; but what is missed is that such embodiment relations may take any number of directions. Both the sports car driver within the constraints of the racing route and the bulldozer driver destroying a rainforest may have the satisfactions of powerful embodiment relations.

There is also a deeper desire which can arise from the experience of embodiment relations. It is the doubled desire that, on one side, is a wish for *total transparency*, total embodiment, for the technology to truly "become me." Were this possible, it would be equivalent to there being no technology, for total transparency would *be* my body and senses; I desire the face-to-face that I would experience without the technology. But that is only one side of the desire. The other side is the desire to have the power, the transformation that the technology makes available. Only by using the technology is my bodily power enhanced and magnified by speed, through distance, or by any of the other ways in which technologies change my capacities. These capacities are always *different* from my naked capacities. The desire is, at best, contradictory. I want the transformation that the technology allows, but I want it in such a way that I am basically unaware of its presence. I want it in such a way that it becomes me. Such a desire both secretly *rejects* what technologies are and overlooks the transformational effects which are necessarily tied to human-technology relations. This illusory desire belongs equally to pro- and anti-technology interpretations of technology.

The desire is the source of both utopian and dystopian dreams. The actual, or material, technology always carries with it only a partial or quasi-transparency, which is the price for the extension of magnification that technologies give. In extending bodily capacities, the technology also transforms them. In that sense, all technologies in use are non-neutral. They change the basic situation, however subtly, however minimally; but this is the other side of the desire. The desire is simultaneously a desire for a change in situation—to inhabit the earth, or even to go beyond the earth—while sometimes inconsistently and secretly wishing that this movement could be without the mediation of the technology.

The direction of desire opened by embodied technologies also has its positive and negative thrusts. Instrumentation in the knowledge activities, notably science, is the gradual extension of perception into new realms. The desire is to see, but seeing is seeing through instrumentation. Negatively, the desire for pure transparency is the wish to escape the limitations of the material technology. It is a platonism returned in a new form, the desire to escape the newly extended body

of technological engagement. In the wish there remains the contradiction: the user both wants and does not want the technology. The user
wants what the technology gives but does not want the limits, the
transformations that a technologically extended body implies. There is
a fundamental ambivalence toward the very human creation of our
own earthly tools.

The ambivalence that can arise concerning technics is a reflection
of one kind upon the *essential ambiguity* that belongs to technologies
in use. But this ambiguity, I shall argue, has its own distinctive shape.
Embodiment relations display an essential magnification/reduction
structure which has been suggested in the instrumentation examples.
Embodiment relations simultaneously magnify or amplify and reduce or
place aside what is experienced through them.

The sight of the mountains of the moon, through all the transformational power of the telescope, removes the moon from its setting in
the expanse of the heavens. But if our technologies were only to repli-
cate our immediate and bodily experience, they would be of little use
and ultimately of little interest. A few absurd examples might show
this:

In a humorous story, a professor bursts into his club with the an-
nouncement that he has just invented a reading machine. The machine scans the pages, reads them, and perfectly reproduces them.
(The story apparently was written before the invention of photocopy-
ing. Such machines might be said to be "perfect reading machines" in
actuality.) The problem, as the innocent could see, was that this machine leaves us with precisely the problem we had prior to its invention. To have reproduced through mechanical "reading" all the books
in the world leaves us merely in the library.

A variant upon the emperor's invisible clothing might work as
well. Imagine the invention of perfectly transparent clothing through
which we might technologically experience the world. We could see
through it, breathe through it, smell and hear through it, touch
through it. Indeed, it effects no changes of any kind, since it is *perfectly* invisible. Who would bother to pick up such clothing (even if
the presumptive wearer could find it)? Only by losing some invisibility—say, with translucent coloring—would the garment begin to be
usable and interesting. For here, at least, fashion would have been invented—but at the price of losing total transparency—by becoming
that through which we relate to an environment.

Such stories belong to the extrapolated imagination of fiction,
which stands in contrast to even the most minimal actual embodiment
relations, which in their material dimensions simultaneously extend
and reduce, reveal and conceal.

In actual human-technology relations of the embodiment sort, the
transformational structures may also be exemplified by variations: In
optical technologies, I have already pointed out how spatial significa-

tions change in observations through lenses. The entire gestalt changes. When the apparent size of the moon changes, along with it the apparent position of the observer changes. Relativistically, the moon is brought "close"; and equivalently, this optical near-distance applies to both the moon's appearance and my bodily sense of position. More subtly, every dimension of spatial signification also changes. For example, with higher and higher magnification, the well-known phenomenon of depth, instrumentally mediated as a "focal plane," also changes. Depth diminishes in optical near-distance.

A related phenomenon in the use of an optical instrument is that it transforms the spatial significations of vision in an instrumentally focal way. But my seeing without instrumentation is a full bodily seeing—I see not just with my eyes but with my whole body in a unified sensory experience of things. In part, this is why there is a noticeable irreality to the apparent position of the observer, which only diminishes with the habits acquired through practice with the instrument. But the optical instrument cannot so easily transform the entire sensory gestalt. The focal sense that is magnified through the instrument is monodimensioned.

Here may be the occasion (although I am not claiming a cause) for a certain interpretation of the senses. Historians of perception have noted that, in medieval times, not only was vision not the supreme sense but sound and smell may have had greatly enhanced roles so far as the interpretation of the senses went. Yet in the Renaissance and even more exaggeratedly in the Enlightenment, there occurred the reduction to sight as the favored sense, and within sight, a certain reduction of sight. This favoritism, however, also carried implications for the other senses.

One of these implications was that each of the senses was interpreted to be clear and distinct from the others, with only certain features recognizable through a given sense. Such an interpretation impeded early studies in echo location.

In 1799 Lazzaro Spallanzani was experimenting with bats. He noticed not only that they could locate food targets in the dark but also that they could do so blindfolded. Spallanzani wondered if bats could guide themselves by their ears rather than by their eyes. Further experimentation, in which the bats' ears were filled with wax, showed that indeed they could not guide themselves without their ears. Spallanzani surmised that either bats locate objects through hearing or they had some sense of which humans knew nothing. Given the doctrine of separate senses and the identification of shapes and objects through vision alone, George Montagu and Georges Cuvier virtually laughed Spallanzani out of the profession.

This is not to suggest that such an interpretation of sensory distinction was due simply to familiarity with optical technologies, but the common experience of enhanced vision through such technologies was at least the standard practice of the time. Auditory technologies were to come later. When auditory technologies did become common, it was possible to detect the same amplification/reduction structure of the human-technology experience.

The telephone in use falls into an auditory embodiment relation. If the technology is good, I hear *you* through the telephone and the apparatus "withdraws" into the enabling background:

(I-telephone)-you

But as a monosensory instrument, your phenomenal presence is that of a voice. The ordinary multidimensioned presence of a face-to-face encounter does not occur, and I must at best imagine those dimensions through your vocal gestures. Also, as with the telescope, the spatial significations are changed. There is here an auditory version of visual near-distance. It makes little difference whether you are geographically near or far, none at all whether you are north or south, and none with respect to anything but your bodily relation to the instrument. Your voice retains its partly irreal near-distance, reduced from the full dimensionality of direct perceptual situations. This telephonic distance is different both from immediate face-to-face encounters and from visual or geographical distance as normally taken. Its distance is a mediated distance with its own identifiable significations.

While my primary set of variations is to locate and demonstrate the invariance of a magnification/reduction structure to any embodiment relation, there are also secondary and important effects noted in the histories of technology. In the very first use of the telephone, the users were fascinated and intrigued by its auditory transparency. Watson heard and recognized Bell's *voice*, even though the instrument had a high ratio of noise to message. In short, the fascination attaches to magnification, amplification, enhancement. But, contrarily, there can be a kind of forgetfulness that equally attaches to the reduction. What is *revealed* is what excites; what is concealed may be forgotten. Here lies one secret for technological trajectories with respect to development. There are *latent telics* that occur through inventions.

Such telics are clear enough in the history of optics. Magnification provided the fascination. Although there were stretches of time with little technical progress, this fascination emerged from time to time to have led to compound lenses by Galileo's day. If some magnification shows the new, opens to what was poorly or not at all previously detected, what can greater magnification do? In our own time, the explosion of such variants upon magnification is dramatic. Electron enhancement, computer image enhancement, CAT and NMR internal scanning, "big-eye" telescopes—the list of contemporary magnificational and visual instruments is very long.

I am here restricting myself to what may be called a *horizontal* trajectory, that is, optical technologies that bring various micro- or macro-phenomena to vision through embodiment relations. By restricting examples to such phenomena, one structural aspect of embodiment relations may be pointed to concerning the relation to microperception and its Adamic context. While *what* can be seen has changed dramatically—Galileo's New World has now been enhanced by astronomical phenomena never suspected and by micro-phenomena still being discovered—there remains a strong phenomenological constant in *how* things are seen. All lenses and optical technologies of the sort being described bring what is to be seen into a normal bodily space and distance. Both the macroscopic and the microscopic appear within the same near-distance. The "image size" of galaxy or amoeba is the *same*. Such is the existential condition for visibility, the counterpart to the technical condition, that the instrument makes things visually present.

The mediated presence, however, must fit, be made close to my actual bodily position and sight. Thus there is a reference within the instrumental context to my face-to-face capacities. These remain primitive and central within the new mediational context. Phenomenological theory claims that for every change in what is seen (the object correlate), there is a noticeable change in how (the experiential correlate) the thing is seen.

In embodiment relations, such changes retain both an equivalence and a difference from non-mediated situations. What remains constant is the bodily focus, the reflexive reference back to my bodily capacities. What is seen must be seen from or within my visual field, from the apparent distance in which discrimination can occur regarding depth, etc., just as in face to face relations. But the range of what can be brought into this proximity is transformed by means of the instrument.

Let us imagine for a moment what was never in fact a problem for the history of instrumentation: If the "image size" of both a galaxy and an amoeba is the "same" for the observer using the instrument, how can we tell that one is macrocosmic and the other microcosmic? The "distance" between us and these two magnitudes, Pascal noted, was the same in that humans were interpreted to be between the infinitely large and the infinitely small.

What occurs through the mediation is not a problem *because our construction of the observation presupposes ordinary praxical spatiality.* We handle the paramecium, placing it on the slide and then under the microscope. We aim the telescope at the indicated place in the sky and, before looking through it, note that the distance is at least that of the heavenly dome. But in our imagination experiment, what if our human were *totally immersed* in a technologically mediated world? What if, from birth, all vision occurred only through lens systems? Here the problem would become more difficult. But in our distance from Adam, it is precisely the presumed difference that makes it possible for us to see both nakedly *and* mediately—and thus to be able to locate the difference—that places us even more distantly from any Garden. It is because we retain this ordinary spatiality that we have a reflexive point of reference from which to make our judgments.

The noetic or bodily reflexivity implied in all vision also may be noticed in a magnified way in the learning period of embodiment. Galileo's telescope had a small field, which, combined with early hand-held positioning, made it very difficult to locate any particular phenomenon. What must have been noted, however, even if not commented upon, was the exaggerated sense of bodily motion experienced through trying to fix upon a heavenly body—and more, one quickly learns something about the earth's very motion in the attempt to use such primitive telescopes. Despite the apparent fixity of the stars, the hand-held telescope shows the earth-sky motion dramatically. This magnification effect is within the experience of one's own bodily viewing.

This bodily and actional point of reference retains a certain privilege. All experience refers to it in a taken-for-granted and recoverable way. The bodily condition of the possibility for seeing is now twice indicated by the very situation in which mediated experience occurs. Embodiment relations continue to locate that privilege of my being here. The partial symbiosis that occurs in well-designed embodied technologies retains that motility which can be called expressive. Embodiment relations constitute one existential form of the full range of the human-technology field.

B. HERMENEUTIC TECHNICS

Heidegger's hammer in use displays an embodiment relation. Bodily action through it occurs within the environment. But broken, missing, or malfunctioning, it ceases to be the means of praxis and becomes an obtruding *object* defeating the work project. Unfortunately, that negative derivation of objectness by Heidegger carries with it a block against understanding a second existential human-technology relation, the type of relation I shall term *hermeneutic*.

The term hermeneutic has a long history. In its broadest and simplest sense it means "interpretation," but in a more specialized sense it refers to *textual* interpretation and thus entails *reading*. I shall retain both these senses and take hermeneutic to mean a special interpretive action within the technological context. That kind of activity calls for special modes of action and perception, modes analogous to the reading process.

Reading is, of course, a reading of _____ ; and in its ordinary context, what fills the intentional blank is a text, something *written*. But all writing entails technologies. Writing has a product. Historically, and more ancient than the revolution brought about by such crucial technologies as the clock or the compass, the invention and development of writing was surely even more revolutionary than clock or compass with respect to human experience. Writing transformed the very perception and understanding we have of language. Writing is a techno- logically embedded form of language.

There is a currently fashionable debate about the relationship between speech and writing, particularly within current Continental philosophy. The one side argues that speech is primary, both historically and ontologically, and the other—the French School—inverts this relation and argues for the primacy of writing. I need not enter this debate here in order to note the *technological difference* that obtains between oral speech and the materially connected process of writing, at least in its ancient forms.

Writing is inscription and calls for both a process of writing itself, employing a wide range of technologies (from stylus for cuneiform to word processors for the contemporary academic), and other material entities upon which the writing is recorded (from clay tablet to computer printout). Writing is technologically mediated language. From it, several features of hermeneutic technics may be highlighted. I shall take what may at first appear as a detour into a distinctive set of human-technology relations by way of a phenomenology of reading and writing.

Reading is a specialized perceptual activity and praxis. It implicates my body, but in certain distinctive ways. In an ordinary act of reading, particularly of the extended sort, what is read is placed before or somewhat under one's eyes. We read in the immediate context from some miniaturized bird's eye perspective. What is read occupies an expanse within the focal center of vision, and I am ordinarily in a somewhat rested position. If the object-correlate, the "text" in the broadest sense, is a chart, as in the navigational examples, what is represented retains a representational isomorphism with the natural features of the landscape. The chart represents the land- (or sea)scape and insofar as the features are isomorphic, there is a kind of representational "transparency." The chart in a peculiar way "refers" beyond itself to what it represents.

Now, with respect to the embodiment relations previously traced, such an isomorphic representation is both similar and dissimilar to what would be seen on a larger scale from some observation position (at bird's-eye level). It is similar in that the shapes on the chart are reduced representations of distinctive features that can be directly or technologically mediated in face-to-face or embodied perceptions. The reader can compare these similarities. But chart reading is also different in that, during the act of reading, the perceptual focus is the chart itself, a substitute for the landscape.

I have deliberately used the chart-reading example for several purposes. First, the "textual" isomorphism of a representation allows this first example of hermeneutic technics to remain close to yet differenti-

ated from the perceptual isomorphism that occurs in the optical examples. The difference is at least perceptual in that one sees *through* the
optical technology, but now one *sees* the chart as the visual terminus,
the "textual" artifact itself.

Something much more dramatic occurs, however, when the representational isomorphism disappears in a printed text. There is no isomorphism between the printed word and what it "represents,"
although there is some kind of *referential* "transparency" that belongs
to this new technologically embodied form of language. It is apparent
from the chart example that the chart itself becomes the *object of
perception* while simultaneously referring beyond itself to what is not
immediately seen. In the case of the printed text, however, the referential transparency is distinctively different from technologically embodied perceptions. *Textual transparency is hermeneutic transparency,
not perceptual transparency.*

Historically, textual transparency was neither immediate nor attained at a stroke. The "technology" of phonetic writing, which now is
increasingly a world-wide standard, became what it is through a series
of variants and a process of experimentation. One early form of writing
was pictographic. The writing was still somewhat like the chart example; the pictograph retained a certain representational isomorphism
with what was represented. Later, more complex ideographic writing
(such as Chinese) was, in effect, a more abstract form of pictography.

Calligraphers have shown that even early phonetic writing followed a gradual process of formalizing and abstracting from a pictographic base (see Figure 4). Letters often depicted a certain animal, the
first syllable of whose name provided the sound for the letter in a simultaneous sound and letter. Built into such early phonetic writing
was thus something like the way the alphabet is still taught to children:
"C is for Cow." Most educated persons are familiar with the mixed
form of writing, hieroglyphics. Although the writing is pictographic, not
all pictographs stood for the entity depicted; some represented sounds
(phonemes).

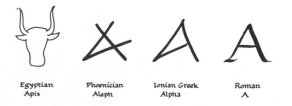

| Egyptian | Phoenician | Ionian Greek | Roman |
| Apis | Aleph | Alpha | A |

An interesting cross-cultural example of this movement from a very
pictographic to a formalized and transformed ideographic writing occurs
with Chinese writing. The same movement from relatively concrete representations in pictographs occurs through abbreviated abstractions—but
in a different direction, non-phonetic and ideographic. Thus, for pho-

netic writing there is a double abstraction (from pictograph to letter and then reconstituting a small finite alphabet into represented spoken words), whereas the doubled abstraction of ideographic writing does not reconstitute to words as such, but to concepts.

In the most ancient Chinese writing in the period of the "Tortoise Shell Language" (prior to 2000 B.C.) and even in some cases through the later "Metal Language" period (2000–500 B.C.), if one is familiar with the objects as they occur within Chinese culture, one can easily detect the pictographic representation involved. For example, one can see in Figure 5 that the ideograph for boat actually abstractly represents the sampan-type boats of the riverways (still in use). Similarly, in the ideograph for gate (Figure 6) one can still recognize the uniquely Orien-tal-type gate in the drawing. The modern variants—related but more abstracted—have clearly lost that instant representational isomorphism.

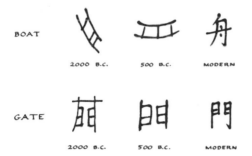

Implied in these transformations are changes of both technique and related technologies. Sergei Eisenstein, the film maker and one sensitive to such image technologies, has pointed to just such a trans-formation which arose out of the invention of the brush and India ink:

> But then, by the end of the third century, the brush is invented. In the first century after the "joyous event" (A.D.)—paper. And, lastly, in the year 220—India ink.
> A complete upheaval. A revolution in draughtmanship. And, after having undergone in the course of history no fewer than fourteen different styles of handwriting, the hieroglyph crystallized in its present form. The means of production (brush and India ink) determined the form.
> The fourteen reforms had their way. As a result:

In the fierily cavorting hieroglyph *ma* (a horse) it is already impossible to recognize the features of the dear little horse sagging pathetically in its hindquarters, in the writing style of Ts'ang Chieh, so well-known from ancient Chinese bronzes.[1]

If this is an accurate portrayal of the evolution of writing, it follows something like a Husserlian origin-of-geometry trajectory. The trajectory was from the more concrete to the greater degrees of abstraction, until virtually all "likeness" to origins disappeared. In this respect, writing only slowly approximated speech.

Once attained, like any other acquisition of the lifeworld, writing could be read and understood in terms of its unique linguistic transparency. Writing becomes an embodied hermeneutic technics. Now the descriptions may take a different shape. What is referred to is referred by the text and is referred to *through* the text. What now presents itself is the "world" of the text.

This is not to deny that all language has its unique kind of transparency. Reference beyond itself, the capacity to let something become present through language, belongs to speech as well. But here the phenomenon being centered upon is the new embodiment of language in writing. Even more thematically, the concern is for the ways in which writing as a "technology" transforms experiential structures.

Linguistic transparency is what makes present the *world* of the text. Thus, when I read Plato, Plato's "world" is made present. But this presence is a *hermeneutic* presence. Not only does it occur *through* reading, but it takes its shape in the interpretative context of my language abilities. His world is linguistically mediated, and while the words may elicit all sorts of imaginative and perceptual phenomena, it is through language that such phenomena occur. And while such phenomena may be strikingly rich, they do not appear *as* word-like.

We take this phenomenon of reading for granted. It is a sedimented acquisition of the literate lifeworld and thus goes unnoticed until critical reflection isolates its salient features. It is the same with the wide variety of hermeneutic technics we employ.

The movement from embodiment relations to hermeneutic ones can be very gradual, as in the history of writing, with little-noticed differentiations along the human-technology continuum. A series of wide-ranging variants upon readable technologies will establish the point. First, a fairly explicit example of a readable technology: Imagine sitting inside on a cold day. You look out the window and notice that the snow is blowing, but you are toasty warm in front of the fire. You can clearly "see" the cold in Merleau-Ponty's pregnant sense of perception—but you do not actually *feel* it. Of course, you could, were you to go outside. You would then have a full face-to-face verification of what you had seen.

But you might also see the thermometer nailed to the grape arbor

---

1. Sergei Eisenstein, *Film Form: Essays in Film Theory*, ed. and trans. Jay Leyda (New York: Harcourt, Brace and World, 1949), p. 29.

post and *read* that it is 28°F. You would now "know" how cold it was, but you still would not feel it. To retain the full sense of an embodiment relation, there must also be retained some isomorphism with the felt sense of the cold—in this case, tactile—that one would get through face-to-face experience. One could invent such a technology; for example, some conductive material could be placed through the wall so that the negative "heat," which is cold, could be felt by hand. But this is not what the thermometer does.

Instead, you read the thermometer, and in the immediacy of your reading you *hermeneutically* know that it is cold. There is an instantaneity to such reading, as it is an already constituted intuition (in phenomenological terms). But you should not fail to note that *perceptually* what you have seen is the dial and the numbers, the thermometer "text." And that text has hermeneutically delivered its "world" reference, the cold.[2]

Such constituted immediacy is not always available. For instance, although I have often enough lived in countries where Centigrade replaces Fahrenheit, I still must translate from my intuitive familiar language to the less familiar one in a deliberate and self-conscious hermeneutic act. Immediacy, however, is not the test for whether the relation is hermeneutic. A hermeneutic relation mimics sensory perception insofar as it is also a kind of seeing as _____ ; but it is a referential seeing, which has as its immediate perceptual focus seeing the thermometer.

Now let us make the case more complex. In the example cited, the experiencer had both embodiment (seeing the cold) and hermeneutic access to the phenomenon (reading the thermometer). Suppose the house were hermetically sealed, with no windows, and the only access to the weather were through the thermometer (and any other instruments we might include). The hermeneutic character of the relation becomes more obvious. I now clearly have to know how to read the instrumentation and from this reading knowledge get hold of the "world" being referred to.

This example has taken actual shape in nuclear power plants. In the Three Mile Island incident, the nuclear power system was observed only through instrumentation. Part of the delay that caused a near meltdown was *misreadings* of the instruments. There was no face-to-face, independent access to the pile or to much of the machinery involved, nor could there be.

An intentionality analysis of this situation retains the mediational position of the technology:

I-technology-world
(engineer-instruments-pile)

---

2. This illustration is my version of a similar one developed by Patrick Heelan in his more totally hermeneuticized notion of perception in *Space Perception and the Philosophy of Science* (Berkeley: University of California Press, 1983), p. 193.

The operator has instruments between him or her and the nuclear pile. But—and here, an essential difference emerges between embodiment and hermeneutic relations—what is immediately perceived is the instrument panel itself. It becomes the object of my microperception, although in the special sense of a hermeneutic transparency, I *read* the pile through it. This situation calls for a different formalization:

$$\text{I-(technology-world)}$$

The parenthesis now indicates that the immediate *perceptual* focus of my experience *is* the control panel. I read through it, but this reading is now dependent upon the semi-opaque connection between the instruments and the referent object (the pile). This *connection* may now become enigmatic.

In embodiment relations, what allows the partial symbiosis of myself and the technology is the capacity of the technology to become perceptually transparent. In the optical examples, the glass-maker's and lens-grinder's arts must have accomplished this end if the embodied use is to become possible. Enigmas which may occur regarding embodiment-use transparency thus may occur within the parenthesis of the embodiment relation:

$$\text{(I-technology)} \rightarrow \text{World}$$

enigma position

(This is not to deny that once the transparency is established, thus making microperception clear, the observer may still fail, particularly at the macroperceptual level. For the moment, however, I shall postpone this type of interpretive problem.) It would be an oversimplification of the history of lens-making were not problems of this sort recognized. Galileo's instrument not only was hard to look through but was good only for certain "middle range" sightings in astronomical terms (it did deliver the planets and even some of their satellites). As telescopes became more powerful, levels, problems with chromatic effects, diffraction effects, etc., occurred. As Ian Hacking has noted,

> Magnification is worthless if it magnifies two distinct dots into one big blur. One needs to resolve the dots into two distinct images. . . . It is a matter of diffraction. The most familiar example of diffraction is the fact that shadows of objects with sharp boundaries are fuzzy. This is a consequence of the wave character of light.[3]

Many such examples may be found in the history of optics, technical

---

3. Ian Hacking, *Representing and Intervening* (Cambridge: Cambridge University Press, 1983), p. 195. Hacking develops a very excellent and suggestive history of the use of microscopes. His focus, however, is upon the technical properties that were resolved before microscopes could be useful in the sciences. He and Heelan, however, along with Robert Ackermann, have been among the pioneers dealing with perception and instrumentation in instruments. Cf. also my *Technics and Praxis* (Dordrecht: Reidel Publishers, 1979).

problems that had to be solved before there could be any extended reach within embodiment relations. Indeed, many of the barriers in the development of experimental science can be located in just such limitations in instrumental capacity.

Here, however, the task is to locate a parallel difficulty in the emerging new human-technology relation, hermeneutic relations. The location of the technical problem in hermeneutic relations lies in the *connector* between the instrument and the referent. Perceptually, the user's visual (or other) terminus is *upon* the instrumentation itself. To read an instrument is an analogue to reading a text. But if the text does not correctly refer, its reference object or its world cannot be present. Here is a new location for an enigma:

$$I \longmapsto \text{(technology-world)}$$

enigma position

While breakdown may occur at any part of the relation, in order to bring out the graded distinction emerging between embodiment and hermeneutic relations, a short pathology of connectors might be noted.

If there is nothing that impedes my direct perceptual situation with respect to the instrumentation (in the Three Mile Island example, the lights remain on, etc.), interpretive problems in reading a strangely behaving "text" at least occur in the open; but the technical enigma may also occur within the text-referent relation. How could the operator tell if the instrument was malfunctioning or that to which the instrument refers? Some form of opacity can occur within the technology-referent pole of the relation. If there is some independent way of verifying which aspect is malfunctioning (a return to unmediated face-to-face relations), such a breakdown can be easily detected. Both such occurrences are reasons for instrumental redundancy. But in examples where such independent verification is not possible or untimely, the opacity would remain.

Let us take a simple mechanical connection as a borderline case. In shifting gears on my boat, there is a lever in the cockpit that, when pushed forward, engages the forward gear; upward, neutral; and backwards, reverse. Through it, I can ordinarily feel the gear change in the transmission (embodiment) and recognize the simple hermeneutic signification (forward for forward) as immediately intuitive. Once, however, on coming in to the dock at the end of the season, I disengaged the forward gear—and the propeller continued to drive the boat forward. I quickly reversed—and again the boat continued. The hermeneutic significance had failed; and while I also felt a difference in the way the gear lever felt, I did not discover until later that the clasp that retained the lever itself had corroded, thus preventing any actual shifting at all. But even at this level there can be opacity within the technology-object relation.

The purpose of this somewhat premature pathology of human-technology relations is not to cast a negative light upon hermeneutic relations in contrast to embodiment ones but rather to indicate that there are different locations where perceptual and human-technology relations interact. Normally, when the technologies work, the technology-world relation would retain its unique hermeneutic transparency. But if the I-(technology-world) relation is far enough along the continuum to identify the relation as a hermeneutic one, the intersection of perceptual-bodily relations with the technology changes.

Readable technologies call for the extension of my hermeneutic and "linguistic" capacities *through* the instruments, while the reading itself retains its bodily perceptual location as a relation *with* or *towards* the technology. What is emerging here is the first suggestion of an emergence of the technology as "object" but without its negative Heideggerian connotation. Indeed, the type of special capacity as a "text" is a condition for hermeneutic transparency.

The transformation made possible by the hermeneutic relation is a transformation that occurs precisely through *differences* between the text and what is referred to. What is needed is a particular set of textually clear perceptions that "reduce" to that which is immediately readable. To return to the Three Mile Island example, one problem uncovered was that the instrument panel design was itself faulty. It did not incorporate its dials and gauges in an easily readable way. For example, in airplane instrument panel design, much thought has been given to pattern recognition, which occurs as a perceptual gestalt. Thus, in a four-engined aircraft, the four dials indicating r.p.m. will be coordinated so that a single glance will indicate which, if any, engine is out of synchronization. Such technical design accounts for perceptual structures.

There is a second caution concerning the focus upon connectors and pathology. In all the examples I have used to this point, the hermeneutic technics have involved material connections. (The thermometer employs a physical property of a bimetallic spring or mercury in a column; the instrument panel at TMI employs mechanical, electrical, or other material connections; the shift lever, a simple mechanical connection.) If reading does not employ any such material connections, it might seem that its referentiality is essentially different, yet not even all technological connections are strictly material. Photography retains representational isomorphism with the object, yet does not "materially" connect with its object; it is a minimal beginning of action at a distance.

I have been using contemporary or post-scientific examples, but non-material hermeneutic relations do not obtain only for contemporary humans. As existential relations, they are as "old" as post-Garden humanity. Anthropology and the history of religions have long been familiar with a wide variety of shamanistic praxes which fall into the pattern of hermeneutic technics. In what may at first seem a somewhat outrageous set of examples, note the various "reading" techniques employed in shamanism. The reading of animal entrails, of thrown bones, of bodily marks—all are hermeneutic techniques. The patterns of the entrails, bones, or whatever are taken to *refer* to some state of affairs, instrumentally or textually.

Not only are we here close to a familiar association between magic and the origins of technology suggested by many writers, but we are, in fact, closer to a wider hermeneutic praxis in an intercultural setting. For that reason, the very strangeness of the practice must be critically examined. If the throwing of bones is taken as a "primitive" form of medical diagnosis—which does play a role in shamanism—we might conclude that it is indeed a poor form of hermeneutic relations. What we might miss, however, is that the entire gestalt of what is being diagnosed may differ radically from the other culture and ours.

It may well be that as a focused form of diagnosis upon some particular bodily ailment (appendicitis, for example), the diagnosis will fail. But since one important element in shamanism is a wider diagnosis, used particularly as the occasion of locating certain communal or social problems, it may work better. The sometimes socially contextless emphasis of Western medicine upon a presumably "mechanical" body may overlook precisely the context which the shaman so clearly recognizes. The entire gestalt is different and differently focused, but in both cases there are examples of hermeneutic relations.

In our case, the very success of Western medicine in certain diseases is due to the introduction of technologies into the hermeneutic relation (fever/thermometer; blood pressure/manometer, etc.) The point is that hermeneutic relations are as commonplace in traditional and ancient social groups as in ours, even if they are differently arranged and practiced.

By continuing the intentionality analysis I have been following, one can now see that hermeneutic relations vary the continuum of human-technology-world relations. Hermeneutic relations maintain the general mediation position of technologies within the context of human praxis towards a world, but they also change the variables within the human-technology-world relation. A comparative formalism may be suggestive:

General intentionality relations
Human-technology-world

Variant A: embodiment relations
(I-technology) $\rightarrow$ world

Variant B: hermeneutic relations
I $\rightarrow$ (technology world)

While each component of the relation changes within the correlation, the overall shapes of the variants are distinguishable. Nor are these matters of simply how technologies are experienced.

Another set of examples from the set of optical instruments may illustrate yet another way in which instrumental intentionalities can follow new trajectories. Strictly embodiment relations can be said to work best when there is both a transparency and an isomorphism between perceptual and bodily action within the relation. I have suggested that a trajectory for development in such cases may often be a horizontal one. Such a trajectory not only follows greater and greater degrees of magnification but also entails all the difficulties of a technical nature that go into allowing what is to be seen as though by direct vision. But not all optical technologies follow this strategy. The introduction of hermeneutic possibilities opens the trajectory into what I shall call *vertical* directions, possibilities that rely upon quite deliberate hermeneutic transformations.

It might be said that the telescope and microscope, by extending vision while transforming it, remained *analogue* technologies. The enhancement and magnification made possible by such technologies remain visual and transparent to ordinary vision. The moon remains recognizably the moon, and the microbe—even if its existence was not previously suspected—remains under the microscope a beastie recognized as belonging to the animate continuum. Here, just as the capacity to magnify becomes the foreground phenomenon to the background phenomenon of the reduction necessarily accompanying the magnification, so the similitude of what is seen with ordinary vision remains central to embodiment relations.

Not all optical technologies mediate such perceptions. In gradually moving towards the visual "alphabet" of a hermeneutic relation, deliberate variations may occur which enhance previously undiscernible *differences*:

1) Imagine using spectacles to correct vision, as previously noted. What is wanted is to *return* vision as closely as possible to ordinary perception, not to distort or modify it in any extreme micro- or macroperceptual direction. But now, for snowscapes or sun on the water or desert, we modify the lenses by coloring or polarizing them to cut glare. Such a variation transforms *what* is seen in some degree. Whether we say the polarized lens removes glare or "darkens" the landscape, what is seen is now clearly different from what may be seen through untinted glasses. This difference is a clue which may open a new *telic direction* for development.

2) Now say that somewhere, sometime, someone notes that certain kinds of tinting reveal unexpected results. Such is a much more complex technique now used in infrared satellite photos. (For the moment, I shall ignore the fact that part of this process is a combined embodiment and hermeneutic relation.) If the photo is of the peninsula of Baja California, it will remain recognizable in shape. Geography, what-

ever depth and height representations, etc., remain but vary in a direction different from any ordinary vision. The infrared photo enhances the difference between vegetation and non-vegetation beyond the limits of any isomorphic color photography. This difference corresponds, in the analogue example, to something like a pictograph. It simultaneously leaves certain analogical structures there and begins to modify the representation into a different, non-perceived "representation."

3) Very sophisticated versions of still representative but non-ordinary forms of visual recognition occur in the new heat-sensitive and light-enhanced technologies employed by the military and police. Night scopes which enhance a person's heat radiation still look like a person but with entirely different regions of what stands out and what recedes. In high-altitude observations, "heat shadows" on the ground can indicate an airplane that has recently had its engines running compared to others which have not. Here visual technologies bring into visibility what was not visible, but in a distinctly now perceivable way.

4) If now one takes a much larger step to spectrographic astronomy, one can see the acceleration of this development. The spectrographic picture of a star no longer "resembles" the star at all. There is no point of light, no disk size, no spatial isomorphism at all—merely a band of differently colored rainbow stripes. The naive reader would not know that this was a picture of a star at all—the reader would have to know the language, the alphabet, that has coded the star. The astronomer-hermeneut does know the language and "reads" the visual "ABCs" in such a way that he knows the chemical composition of the star, its internal makeup, rather than its shape or external configuration. We are here in the presence of a more fully hermeneutic relation, the star mediated not only instrumentally but in a transformation such that we must now thematically *read* the result. And only the informed reader can do the reading.

There remains, of course, the *reference* to the star. The spectograph is *of* Rigel or *of* Polaris, but the individuality of the star is now made present hermeneutically. Here we have a beginning of a special transformation of perception, a transformation which deliberately enhances differences rather than similarities in order to get at what was previously unperceived.

5) Yet even the spectrograph is but a more radical transformation of perception. It, too, can be transformed by a yet more radical *hermeneutic* analogue to the *digital* transformation which lies embedded in the preferred quantitative praxis of science. The "alphabet" of science is, of course, mathematics, a mathematics that separates itself by yet another hermeneutic step from perception embodied.

There are many ways in which this transformation can and does occur, most of them interestingly involving a particular act of *translation* that often goes unnoticed. To keep the example as simple as possible, let us assume *mechanical* or *electronic* "translation." Suppose

our spectrograph is read by a machine that yields not a rainbow spectrum but a set of numbers. Here we would arrive at the final hermeneutic accomplishment, the transformation of even the analogue to a
digit. But in the process of hermeneuticization, the "transparency" to
the object referred to becomes itself enigmatic. Here more explicit
and thematic interpretation must occur.

Hermeneutic relations, particularly those utilizing technologies
that permit vertical transformations, move away from perceptual isomorphism. It is the *difference* between what is shown and how something is shown which is informative. In a hermeneutic relation, the
world is first transformed into a text, which in turn is read. There is
potentially as much flexibility within hermeneutic relations as there are
in the various uses of language. Emmanuel Mournier early recognized
just this analogical relationship with language:

> The machine as implement is not a simple material extension of our
> members. It is of another order, an annex to our language, an auxiliary
> language to mathematics, a means of penetrating, dissecting and revealing
> the secret of things, their implicit intentions, their unemployed capacities.[4]

Through hermeneutic relations we can, as it were, *read* ourselves
into any possible situation without being there. In science, in contrast
to literature, what is important is that the reading retain *some* kind of
reference or hermeneutic transparency to what is there. Perhaps that is
one reason for the constant desire to reverse what is read back towards what may be perceived. In this reversal, contemporary technologically embodied science has frequently derived what might be
called *translation technologies*. I mention two in passing:

(a) Digital processes have become *de rigueur* within the perceptual domain. The development of pictures from space probes is such a
double translation process. The photograph of the surface of Venus is a
technological analogue to human vision. It at least is a field display of
the surface, incorporating the various possible figures and contrasts
that would be seen instantaneously in a visual gestalt—but this holistic
result cannot be transmitted in this way by the current technologies.
Thus it is "translated" into a digital code, which can be transmitted.
The "seeing" of the instrument is broken down into a series of digits
that are radiographically transmitted to a receiver; then they are reas-
sembled into a spatter pattern and enhanced to reproduce the photograph taken millions of miles away. It would be virtually impossible for
anyone to read the digits and tell what was to be seen; only when the
linear text of the digits has been retranslated back into the span of an
instantaneous visual gestalt can it be seen that the rocks on Venus are

---

4. Emmanuel Mournier, *Be Not Afraid*, trans. Cynthia Rowland (London: Rockcliffe, 1951), p. 195.

or are not like those on the moon. Here the analogues of perception and language are both utilized to extend vision beyond the earth.

(*b*) The same process is used audially in digital recordings. Once again, the double translation process takes place and sound is reduced to digital form, reproduced through the record, and translated back into an auditory gestalt.

Digital and analogue processes blur together in certain configurations. Photos transmitted as points of black on a white ground and re-assembled within certain size limits are perceptually gestalted; we see Humphrey Bogart, not simply a mosaic of dots. (Pointillism did the same in painting, although in color. So-called concrete poetry employs the same crossover by placing the words of the poem in a visual pattern so the poem may be both read and seen as a visual pattern.)

Such translation and retranslation processes are clearly transformations from perceptually gestalted phenomena into analogues of writing (serial translation and retranslation processes are clearly transformations from perceptual gestalt phenomena into analogues of writing serial transmissions along a "line," as it were), which are then re-translatable into perceptual gestalts.

I have suggested that the movement from embodiment relations to hermeneutic ones occurs along a human-technology continuum. Just as there are complicated, borderline cases along the continuum from fully haired to bald men, there are the same less-than-dramatic differences here. I have highlighted some of this difference by ac-centing the bodily-perceptual distinctions that occur between embodi-ment and hermeneutic relations. This has allowed the difference in perceptual and hermeneutic transparencies to stand out.

There remain two possible confusions that must be clarified be-fore moving to the next step in this phenomenology of technics. First, there is a related sense in which perception and interpretation are intertwined. Perception is primitively already interpretational, in both micro- and macrodimensions. To perceive is already "like" reading. Yet reading is also a specialized act that receives both further defini-tion and elaboration within literate contexts. I have been claiming that one of the distinctive differences between embodiment and herme-neutic relations involves perceptual position, but in the broader sense, interpretation pervades both embodiment and hermeneutic action.

A second and closely related possible confusion entails the double sense in which a technology may be used. It may be used simulta-neously both as something *through* which one experiences and as something *to* which one relates. While this is so, the doubled relation takes shapes in embodiment different from those of hermeneutic rela-tions. Return to the simple embodiment relation illustrated in wearing eyeglasses. *Focally*, my perceptual experience finds its directional aim *through* the lenses, terminating my gaze upon the object of vision; but as a *fringe* phenomenon, I am simultaneously aware of (or can become so) the way my glasses rest upon the bridge of my nose and the tops of my ears. In this fringe sense, I am aware *of* the glasses, but the focal phenomenon is the perceptual transparency that the glasses allow.

In cases of hermeneutic transparency, this doubled role is subtly changed. Now I may carefully read the dials within the core of my visual field and attend to them. But my reading is simultaneously a reading through them, although now the terminus of reference is not necessarily a perceptual object, nor is it, strictly speaking, perceptually present. While the type of transparency is distinct, it remains that the purpose of the reading is to gain hermeneutic transparency.

Both relations, however, at optimum, occur within the familiar acquisitional praxes of the lifeworld. Acute perceptual seeing must be learned and, once acquired, occurs as familiarly as the act of seeing itself. For the accomplished and critical reader, the hermeneutic transparency of some set of instruments is as clear and as immediate as a visual examination of some specimen. The peculiarity of hermeneutic transparency does not lie in either any deliberate or effortful accomplishment of interpretation (although in learning any new text or language, that effort does become apparent). That is why the praxis that grows up within the hermeneutic context retains the same sense of spontaneity that occurs in simple acts of bodily motility. Nevertheless, a more distinctive presence *of* the technology appears in the example. My awareness of the instrument panel is both stronger and centered more focally than the fringe awareness of my eyeglasses frames, and this more distinct awareness is essential to the optimal use of the instrumentation.

In both embodiment and hermeneutic relations, however, the technology remains short of full objectiveness or *otherwise*. It remains the means through which something else is made present. The negative characterization that may occur in breakdown pathologies may return. When the technology in embodiment position breaks down or when the instrumentation in hermeneutic position fails, what remains is an obtruding, and thus negatively derived, object.

Both embodiment and hermeneutic relations, while now distinguished, remain basic existential relations between the human user and the world. There is the danger that my now-constant and selective use of scientific instrumentation could distort the full impact of the existential dimension. Prior to moving further along the human-technology-world continuum, I shall briefly examine a very different set of instrumental examples. The instrumentation in this case will be *musical instrumentation*.

In the most general sense, it should be easy to see that the use of musical instrumentation, in performance, falls into the same configurations as do scientific instruments:

I—musical instrument—world
I—scientific instrument—world

But the praxical context is significantly changed. If scientific or knowledge-developing praxis is constrained by the need to have a referential terminus within the world, the musical praxis is not so constrained. Indeed, if there is a terminus, it is a reference not so much to some thing or region of the environment as to the production of a musical event within that environment. The "musical object" is whatever sound phenomenon occurs through the performance upon the instrument. Musical sounds are produced, *created*. Whereas in the development of scientific instrumentation the avoidance of phenomena that would be artifacts of the instrument rather than of its referent are to be avoided or reduced as much as possible, the very discovery and enhancement of such instrumental artifacts may be a positive phenomenon in making music. There are interesting and significant differences in these two praxical contexts, but for the moment, I shall restrict myself to a set of observations about the similarities in the intentionality structures of both scientific and musical instrumentation.

It should be obvious that a very large use of musical instrumentation falls clearly into the embodiment relation pattern. The player picks up the instrument (having learned to embody it) and expressively produces the desired music:

Player-instrument-sound

In embodiment cases, the sound-making instrument will be partially symbiotically embodied:

(Player-instrument)-sound

Second, the previously noted amplification/reduction structure also occurs here. If our player is a trombonist, the "buzz" his lip vibrations produce can be heard without any instrument but, once amplified and transformed *through* the trombone, occur as the musical sound distinctive to the human-instrument pairing. Equally immediately, at least within the complex of contemporary instrumentation, one may detect that nothing like a restriction to human sound as such belongs to the contemporary musical context. Isomorphism to human sound, while historically playing a significant cultural role, now occupies only one dimension of musical sound.

This history, however, is interesting. There have been tendencies in Western musical history to restrict to or at least to develop precisely along horizontally variant ways. The restriction of musical sound to actual human voices (certain Mennonite sects do not allow any musical instrumentation, and all hymn singing is done a cappella) is a form of this tendency. Instrumentation that mimics or actually amplifies vocal sounds and their ranges is another example: woodwinds, horns, organs (even to the organ stop titles which are usually voice analogues)—all are ancient instruments that often deliberately followed a kind of vocal isomorphism. Medieval music was often doubly constrained. Not only must the music remain within the range of human similitude, but even the normatively controlled harmonics and chant lines were religio-culturally constrained. Later, one could detect a much more vocal

model to much Italian (Renaissance through Baroque) music in contrast to a more instrumentally oriented model in German music.

The implicit valuational model of the human voice was also reflected in the music history of the West by the ranking of instruments
by *expressivity*, with those instruments thought most expressive—the
violin, for example—rated more highly than those farther from the vocal model.

The difference between embodiment and hermeneutic relations
appears within this context as well. While embodiment relations in the
most general existential sense need not be strictly constrained by isomorphism, hermeneutic variants occur very quickly along the musical
spectrum. The piano retains little vocal isomorphism; yet when played,
it falls into the embodiment relation, is expressive of the individual
style and attainment of the performer, etc. Farther along the continuum, computer-produced music clearly occurs much more fully within
the range of hermeneutic relations, in some cases with the emergence
of random-sound generation very close to the sense of *otherness*,
which will characterize the next set of relations where the technology
emerges as *other*.

Instrumental music, as technics, may go in either embodiment or
hermeneutic directions. It may develop its instrumentation in both vertical and horizontal trajectories. In either direction there are recognizable clear, technological transformations. If the Western "bionic"
model of much early music was voice, in Andean music it was bird
song (both in melody and in sound quality produced by breathy wood
flutes). Contrarily, percussion instrumentation (drum music and communication) was, from the outset, a movement in a vertical and thus
more hermeneutic direction. This exploration of possibility trees in
horizontal and vertical directions belongs to the realm of musical
praxis as much as to scientific, but is without any referentiality to a
natural world.

The result of technological development in musical technics is
also suggestively different from its result in scientific praxis. The
"world" produced musically through all the technical adumbrations is
not that suggested either by the new philosophy of science or by a
Heideggerian philosophy of technology. The closest analogy to the notion of standing reserve (resource well) that the musical "world" might
take is that the realm of all possible sound may be taken and/or transformed musically. But the acoustical resources of musical technics are
utilized through the creative sense of *play* which pervades musical
praxis. The "musical object" is a created object, but its creation is not
constrained by the same imperatives of scientific praxis. Yet the materialization of musical sound *through* instrumentation remains a fully
human technological form of action.

What can be glimpsed in this detour into musical instrumentation
is that while the human-technology structures are parallel with those

found within scientific instrumentation, the "world" created does not at all imply the same reduction to what has been claimed as the unique Western view of the domination of nature. Here, then, is an opening to a different possible trajectory of development.

C. ALTERITY RELATIONS

Beyond hermeneutic relations there lie *alterity relations*. The first suggestions of such relations, which I shall characterize as relations *to* or *with* a technology, have already been suggested in different ways from within the embodiment and hermeneutic contexts. Within embodiment relations, were the technology to intrude upon rather than facilitate one's perceptual and bodily extension into the world, the technology's objectness would necessarily have appeared negatively. Within hermeneutic relations, however, there emerged a certain positivity to the objectness of instrumental technologies. The bodily-perceptual focus *upon* the instrumental text is a condition of its own peculiar hermeneutic transparency. But what of a positive or presentential sense of relations with technologies? In what phenomenological senses can a technology be *other*?

The analysis here may seem strange to anyone limited to the habits of objectivist accounts, for in such accounts technologies as objects usually come first rather than last. The problem for a phenomenological account is that objectivist ones are non-relativistic and thus miss or submerge what is distinctive about human-technology relations.

A naive objectivist account would likely begin with some attempt to circumscribe or define technologies by object characteristics. Then, what I have called the technical properties of technologies would become focal. Some combination of physical and material properties would be taken to be definitional. (This is an inherent tendency of the standard nomological positions such as those of Bunge and Hacking). The definition will often serve a secondary purpose by being stipulative: only those technologies that are obviously dependent upon or strongly related to contemporary scientific and industrial productive practices will count.

This is not to deny that objectivist accounts have their own distinctive strengths. For example, many such accounts recognize that technological or "artificial" products are different from the simply found object or the natural object. But the submergence of the human-technology relation remains hidden, since either object may enter into praxis and both will have their material, and thus limited, range of technical usability within the relation. Nor is this to deny that the objectivist accounts of types of technologies, types of organization, or types of designed purposes should be considered. But the focus in this first program remains the phenomenological derivation of the set of human-technology relations.

There is a tactic behind my placing alterity relations last in the order of focal human-technology relations. The tactic is designed, on the one side, to circumvent the tendency succumbed to by Heidegger and his more orthodox followers to see the otherness of technology only in negative terms or through negative derivations. The hammer example, which remains paradigmatic for this approach, is one that derives objectness from breakdown. The broken or missing or malfunctioning technology could be *discarded*. From being an obtrusion it could become *junk*. Its objectness would be clear—but only partly so. Junk is not a focal object of use relations (except in certain limited situations). It is more ordinarily a background phenomenon, that which has been put out of use.

Nor, on the other side, do I wish to fall into a naively objectivist account that would simply concentrate upon the material properties of the technology as an object of knowledge. Such an account would submerge the relativity of the intentionality analysis, which I wish to preserve here. What is needed is an analysis of the positive or presentential senses in which humans relate to technologies as relations *to* or with technologies, to technology-as-other. It is this sense which is included in the term "alterity."

Philosophically, the term "alterity" is borrowed from Emmanuel Levinas. Although Levinas stands within the traditions of phenomenology and hermeneutics, his distinctive work, *Totality and Infinity*, was "anti-Heideggerian." In that work, the term "alterity" came to mean the radical difference posed to any human by another human, an *other* (and by the ultimately other, God). Extrapolating radically from within the tradition's emphasis upon the non-reducibility of the human to either objectness (in epistemology) or as a means (in ethics), Levinas poses the otherness of humans as a kind of *infinite* difference that is concretely expressed in an ethical, face-to-face encounter.

I shall retain but modify this radical Levinasian sense of human otherness in returning to an analysis of human-technology relations. How and to what extent do technologies become other or, at least, *quasi-other*? At the heart of this question lie a whole series of well-recognized but problematic interpretations of technologies. On the one side lies the familiar problem of anthropomorphism, the personalization of artifacts. This range of anthropomorphism can reach from serious artifact-human analogues to trivial and harmless affections for artifacts.

An instance of the former lies embedded in much AI research. To characterize computer "intelligence" as human-like is to fall into a peculiarly contemporary species of anthropomorphism, however sophisticated. An instance of the latter is to find oneself "fond" of some particular technofact as, for instance, a long-cared-for automobile which one wishes to keep going and which may be characterized by quite deliberate anthropomorphic terms. Similarly, in ancient or non-

Western cultures, the role of sacredness attributed to artifacts exemplifies another form of this phenomenon.

The religious object (idol) does not simply "represent" some absent power but is endowed with the sacred. Its aura of sacredness is spatially and temporally present within the range of its efficacy. The tribal devotee will defend, sacrifice to, and care for the sacred artifact. Each of these illustrations contains the seeds of an alterity relation.

A less direct approach to what is distinctive in human-technology alterity relations may perhaps better open the way to a phenomenologically relativistic analysis. My first example comes from a comparison to a technology and to an animal "used" in some practical (although possibly sporting) context: the spirited horse and the spirited sports car.

To ride a spirited horse is to encounter a lively animal *other.* In its pre- or nonhuman context, the horse has a life of its own within the environment that allowed this form of life. Once domesticated, the horse can be "used" as an "instrument" of human praxis—but only to a degree and in a way different from counterpart technologies; in this case, the "spirited" sports car.

There are, of course, analogues which may at first stand out. Both horse and car give the rider/driver a magnified sense of power. The speed and the experience of speed attained in riding/driving are dramatic extensions of my own capacities. Some prominent features of embodiment relations can be found analogously in riding/driving. I experience the trail/road through horse/car and guide/steer the mediating entity under way. But there are equally prominent differences. No matter how well trained, no horse displays the same "obedience" as the car. Take malfunction: in the car, a malfunction "resists" my command—I push the accelerator, and because of a clogged gas line, there is not the response I expected. But the animate resistance of a spirited horse is more than such a mechanical lack of response—the response is more than malfunction, it is *disobedience.* (Most experienced riders, in fact, prefer spirited horses over the more passive ones, which might more nearly approximate a mechanical obedience.) This life of the other in a horse may be carried much further—it may live without me in the proper environment; it does not need the *deistic* intervention of turning the starter to be "animated." The car will not shy at the rabbit springing up in the path any more than most horses will obey the "command" of the driver to hit the stone wall when he is too drunk to notice. The horse, while approximating some features of a mediated embodiment situation, never fully enters such a relation in the way a technology does. Nor does the car ever attain the sense of animation to be found in horseback riding. Yet the analogy is so deeply embedded in our contemporary consciousness (and perhaps the lack of sufficient experience with horses helps) that we might be tempted to emphasize the similarities rather than the differences.

Anthropomorphism regarding the technology on the one side and the contrast with horseback riding on the other point to a first approximation to the unique type of otherness that relations to technologies hold. Technological otherness is a *quasi-otherness*, stronger than mere objectness but weaker than the otherness found within the animal kingdom or the human one; but the phenomenological derivation must center upon the positive experiential aspects outlining this relation.

In yet another familiar phenomenon, we experience technologies as *toys* from childhood. A widely cross-cultural example is the spinning top. Prior to being put into use, the top may appear as a top-heavy object with a certain symmetry of design (even early tops approximate the more purely functional designs of streamlining, etc.), but once "deistically" animated through either stick motion or a string spring, the now spinning top appears to take on a life of its own. On its tip (or "foot") the top appears to defy its top-heaviness and gravity itself. It traces unpredictable patterns along its pathway. It is an object of *fascination*.

Note that once the top has been set to spinning, what was imparted through an embodiment relation now exceeds it. What makes it fascinating is this property of quasi-animation, the life of its own. Also, of course, once "automatic" in its motion, the top's movements may be entered into a whole series of possible contexts. I might enter a game of warring tops in which mine (suitably marked) represents me. If I-as-top am successful in knocking down the other tops, then this game of hermeneutics has the top winning for me. Similarly, if I take its quasi-autonomous motion to be a hermeneutic predictor, I may enter a divination context in which the path traced or the eventual point of stoppage indicates some fortune. Or, entering the region of scientific instrumentation, I may transform the top into a gyroscope, using its constancy of direction within its now-controlled confines as a better-than-magnetic compass. But in each of these cases, the top may become the focal center of attention as a quasi-other to which I may relate. Nor need the object of fascination carry either an embodiment or hermeneutic referential transparency.

To the ancient and contemporary top, compare briefly the fascination that occurs around video games. In the actual use of video games, of course, the embodiment and hermeneutic relational dimensions are present. The joystick that embodies hand and eye coordination skills extends the player into the displayed field. The field itself displays some hermeneutic context (usually either some "invader" mini-world or some sports analogue), but this context does not refer beyond itself into a worldly reference.

In addition to these dimensions, however, there is the sense of *interacting with* something other than me, the technological *competitor*. In competition there is a kind of dialogue or exchange. It is the quasi-animation, the quasi-otherness of the technology that fascinates and challenges. I must beat the machine or it will beat me.

In each of the cases mentioned, features of technological alterity have shown themselves. The quasi-otherness, the quasi-autonomy which appears in the toy or the game is a variant upon the technologies that have fascinated Western thinkers for centuries, the *automaton*.

The most sophisticated Greek (and similarly, Chinese) technologies did not appear in practical or scientific contexts so often as in game or theatrical ones. (War contexts, of course, have always employed advanced technologies.) Within these contexts, automatons were devised. From rediscovered treatises by Hero of Alexandria on pneumatics and hydraulics (which had in the second century B.C. already been used for humorous applications), the Renaissance builders began to construct various automata. The applications of Hero had been things like automatically opening temple doors and artificial birds that sang through steam whistles. In the Renaissance reconstructions, automata became more complex, particularly in fountain systems:

> The water garden of the Villa d'Este, built in 1550 at Tivoli, outside Rome, for the son of Lucrezia Borgia [was the best known]. The slope of the hill was used to supply fountains and dozens of grottos where water-powered figures moved and played and spouted. . . . The Chateau Merveilleux of Helbrun . . . is full of performing figures of men and women where fountains turn on and off unexpectedly or, operating in the intricate and quite amazing theatre of puppets, run by water power.[5]

The rage for automata was later to develop in a number of directions from music machines, of which the Deutsches Museum in Munich has a grand collection, to Vaucanson's automated duck which quacked, ate, drank, and excreted.[6] Much later, automation techniques were used in more practical contexts, although versions of partially automated looms for textiles did begin to appear in the eighteenth century (Vaucanson, the maker of the automated duck, invented the holed cylinder that preceded the punch-card system of the Jacquard loom).

Nor should the clock be exempted from this glance at automata fascination. The movements of the heavens, of the march of life and death, and of the animated figures on the clocks of Europe were other objects of fascination that seemed to move "autonomously." The superficial aspects of automation, the semblance of the animate and the similitude of the human and animal, remained the focus for even more serious concerns with automatons. That which is more "like" us seemed to center the fascination and make the alterity more quasi-animate.

---

5. Burke, *Connections*, p. 106.
6. Ibid., p. 107.

Fascination may hide what is reductive in technological selectivities. But it may also hide, doubly, a second dimension of an instrumental intentionality, its possible dissimilarity direction, which may often prove in the longer run the more interesting trajectory of development. Yet semblance usually appears to be the first focus.

It was this *semblance* which became a worry for Modern (seventeenth and eighteenth century) Philosophy. Descartes's famous doubts also utilize the popular penchant for automata. In seeking to prove that it is the mind alone and not the eyes that know things, he argues:

> I should forthwith be disposed to conclude that the wax is known by the act of sight and not by the intuition of the mind alone were it not for the analogous instance of human beings passing on in the street below, as observed from a window. In this case, I do not fail to say that I see the men themselves, just as I say that I see the wax; and yet, what do I see from the window beyond hats and cloaks that might cover artificial machines, whose motions might be determined by springs?[7]

This can-I-be-fooled-by-a-cleverly-conceived-robot argument was to have an exceedingly long history, even into the precincts of contemporary analytic philosophies.

Were Descartes to become a contemporary of current developments in the attempt to mimic animal and human motions by automata, he might well rethink his illustration. Not only spring-run automata but also the most sophisticated computer-run automata look mechanical. These most sophisticated computer-run automata have difficulty maneuvering in anything like a lifelike motion. As Dreyfus has pointed out and as would be confirmed by many current researchers, bodily motion is perhaps harder to imitate than certain "mental" activities such as calculating.

To follow only the inclination towards similitude, however, is to reduce what may be learned from our relations with technologies. The current state of the art in AI research, for example, while having been partially freed from its earlier fundamentalistic state, remains primarily within the aim of creating similarities with human intelligence or modeling what are believed to be analogues to our intelligence. Yet it might well be that the *differences* that emerge from computer experimentation may be more informative or, at least, as informative as the similitudes.

There are what I shall call technological *intentionalities* that emerge from many technologies. Let us engage in a pseudo-Cartesian, imaginative construction of a humanoid robot, within the limits of easily combinable and available technologies, to take account of the similarity/difference structures which may be displayed. I shall begin with

---

7. René Descartes, *A Discourse on Method*, trans. John Veitch (London: J. M. Dent, 1953), p. 92.

the technology's "perceptions" of sensory equipment: What if the robot were to hear? The inventor, perhaps limited by a humanist's budget, could install an omnidirectional microphone for ears. We could check upon what our robot would "hear" by adding a cassette player for a recorded "memory" of its "hearing." What is heard would turn out to be very differently structured, to have a very different form of intentionality than what any human listener would hear.

Assume that our robot is attending a university lecture in a large hall and is seated, as a shy student might be, near the rear. Given the limits of the mentioned technology, what would be heard would fail to have either the foreground/background pattern of human listening or the selective elimination of noise that even ordinary listening displays. The robot's auditory memory, played back, would reveal something much more like a sense-data auditory world than the one we are familiar with. The lecturer's voice, though recorded and within low limits perhaps detectable, would often be buried under the noise and background sounds that are selectively masked by human listening. For other purposes, precisely this differently structured technological intentionality could well be useful and informative. Such a different auditory selectivity could perhaps give clues to better architectural dampening of sounds precisely because what is repressed in human listening here stands out. In short, there is "truth" to be found in both the similarity and the difference that technological intentionalities reveal.

A similar effect could be noted with respect to the robot's vision. Were its eyes to be made of television equipment and the record or memory of what it has seen displayed on a screen, we would once again note the flatness of its visual field. Depth phenomena would be greatly reduced or would disappear. Although we have become accustomed to this flat field in watching television, it is easy to become reaware of the lack of depth between the baseball pitcher and the batter upon the screen. The technological shape of intentionality differs significantly from its human counterpart.

The fascination with human or animate similitude within the realm of alterity relations is but another instance of the types of fascination pervading our relations with technologies. The astonishment of Galileo at what he saw through the telescope was, in effect, the location of similitude within embodiment use. The magnification was the magnification *of* human visual capacity and remained within the range of what was familiarly visible. The horizontal trajectory of magnification that can more and more enhance vision is a trajectory along an already familiar praxis.

With the examples of fascination with automata, the fascination also remains within the realm of the familiar, now in a kind of mirror phenomenon for humans and the technology. Of all the animals in the earth's realm, it seems that the human ones are those who can pro-

long this fascination the most intensely. Paul Levinson, in an examination of the history of media technologies, has argued that there are
three stages through which technologies pass. The first is that of technology as toy or novelty. The history of film technology is instructive:

> The first film makers were not artists but tinkers. . . . "Their goal in making
> a movie was not to create beauty but to display a scientific curiosity." A
> survey of the early "talkies" like *The Jazz Singer*, first efforts in animation
> such as Disney's "Laugh-O-Gram" cartoons, and indeed the supposed
> debut of the motion picture in *Fred Ott's Sneeze* supports [this thesis]
> itself.[8]

The same observation could be made about much invention. But once
taken more seriously, novelty can be transformed into a second stage,
according to Levinson: that of technology as mirror of reality. This too
happened in the history of film. Following the early curiosities at the
onset of the film industry, the introduction of the Lumieres' presentation of "actualities" were, in part, fascinating precisely through the
magnification/reduction selectivities that film technologies produce
through unique film intentionalities. Examples could be as mundane as
"workers leaving a factory, a baby's meal, and the famous train entering the station." What made such cinemas vérités dramatic were "in
this case, a real train chugging into a real station, at an angle such that
the audience could almost believe the train was chugging at *them*."[9]

This mirror of life, like the automaton, is not isomorphic with non-
technological experience but is technologically transformed with the
various effects that exaggerate or enhance some effects while simultaneously reducing others. Levinson is quite explicit in his analysis concerning the ways newly introduced technologies also enhance this
development:

> The growth of film from gimmick to replicator was apparently in large part
> dependent upon a new technological component. . . . The "toy" film
> played to individuals who peeked into individual kinetoscopes; but the
> "reality" film reached out to mass audiences, who viewed the reality-
> surrogate in group theatres. The connection between mass audiences and
> reality simulation, moreover, was no accident. Unlike the perception of
> novelties, which is inherently subjective and individualized, reality
> perception is a fundamentally objective, group process.[10]

Although the progression of the analysis here moves from embodiment and hermeneutic relations to alterity ones, the interjection of

8. Paul Levinson, "Toy, Mirror and Art: The Metamorphosis of Technological Culture,"
in *Philosophy, Technology and Human Affairs*, ed. Larry Hickman (College Station: Ibis
Press, 1985), p. 163.
9. Ibid., p. 165.
10. Ibid., p. 167.

film or cinema examples is of suggestive interest. Such technologies are transitional between hermeneutic and alterity phenomena. When I first introduced the notion of hermeneutic relations, I employed what could be called a "static" technology: writing. The long and now ancient technologies of writing result in fixed texts (books, manuscripts, etc., all of which, barring decay or destruction, remain stable in themselves). With film, the "text" remains fixed only in the sense that one can repeat, as with a written text, the seeing and hearing of the cinema text. But the mode of presentation is dramatically different. The "characters" are now animate and theatrical, unlike the fixed alphabetical characters of the written text. The dynamic "world" of the cinema-text, while retaining many of the functional features of writing, also now captures the semblance of real-time, action, etc. It remains to be "read" (viewed and heard), but the object-correlate necessarily appears more "life-like" than its analogue—written text. This factor, naively experienced by the current generations of television addicts, is doubtless one aspect in the problems that emerge between television watching habits and the state of reading skills. James Burke has pointed out that "the majority of the people in the advanced industrialized nations spend more time watching television than doing anything else beside work."[11] The same balance of time use also has shown up in surveys regarding students. The hours spent watching television among college and university students, nationally, are equal to or exceed those spent in doing homework or out-of-class preparation

Film, cinema, or television can, in its hermeneutic dimension, refer in its unique way to a "world." The strong negative response to the Vietnam War was clearly due in part to the virtually unavoidable "presence" of the war in virtually everyone's living room. But films, like readable technologies, are also *presentations*, the focal terminus of a perceptual situation. In that emergent sense, they are more dramatic forms of perceptual immediacy in which the presented display has its own characteristics conveying quasi-alterity. Yet the engagement with the film normally remains short of an engagement with an *other*. Even in the anger that comes through in outrage about civilian atrocities or the pathos experienced in seeing starvation epidemics in Africa, the emotions are not directed to the screen but, indirectly, through it, in more appropriate forms of political or charitable action. To this extent there is retained a hermeneutic reference elsewhere than at the technological instrument. Its quasi-alterity, which is also present, is not fully focal in the case of such media technologies.

A high-technology example of breakdown, however, provides yet another hint at the emergence of alterity phenomena. Word processors have become familiar technologies, often strongly liked by their

---

11. Burke, *Connections*, p. 5.

users (including many philosophers who fondly defend their choices, profess knowledge about the relative abilities of their machines and programs, etc.). Yet in breakdown, this quasi-love relationship reveals its quasi-hate underside as well. Whatever form of "crash" may occur, particularly if some fairly large section of text is involved, it occasions frustration and even rage. Then, too, the programs have their idiosyncrasies, which allow or do not allow certain movements; and another form of human-technology competition may emerge. (Mastery in the highest sense most likely comes from learning to program and thus overwhelm the machine's previous brainpower. "Hacking" becomes the game-like competition in which an entire system is the alterity correlate.) Alterity relations may be noted to emerge in a wide range of computer technologies that, while failing quite strongly to mimic bodily incarnations, nevertheless display a quasi-otherness within the limits of linguistics and, more particularly, of logical behaviors. Ultimately, of course, whatever contest emerges, its sources lie opaquely with other humans as well but also with the transformed technofact, which itself now plays a more obvious role within the overall relational net.

I have suggested that the computer is one of the stronger examples of a technology which may be positioned within alterity relations. But its otherness remains a quasi-otherness, and its genuine usefulness still belongs to the borders of its hermeneutic capacities. Yet in spite of this, the tendency to fantasize its quasi-otherness into an authentic otherness is pervasive. Romanticizations such as the portrayal of the emotive, speaking "Hal" of the movie *2001: A Space Odyssey*, early fears that the "brain power" of computers would soon replace human thinking, fears that political or military decisions will not only be informed by but also made by computers—all are symptoms revolving around the positing of otherness to the technology.

These romanticizations are the alterity counterparts to the previously noted dreams that wish for total embodiment. Were the technofact to be genuinely an other, it would both be and not be a *technology*. But even as quasi-other, the technology falls short of such totalization. It retains its unique role in the human-technology continuum of relations as the medium of transformation, but as a recognizable medium.

The wish-fulfillment desire occasioned by embodiment relations—the desire for a fully transparent technology that would *be* me while at the same time giving me the powers that the use of the technology makes available—here has its counterpart fantasy, and this new fantasy has the same internal contradiction: It both reduces or, here, extrapolates the technology into that which is not a technology (in the first case, the magical transformation is *into me*; in this case, *into the other*), and at the same time, it desires what is not identical with me or the other. The fantasy is for the transformational effects. Both fantasies, in effect, deny technologies playing the roles they do in the human-technology continuum of relations; yet it is only on the condition that there be some detectable differentiation within the relativity that the unique ways in which technologies transform human experience can emerge.

In spite of the temptation to accept the fantasy, what the quasi-otherness of alterity relations does show is that humans may relate positively or presentially *to* technologies. In that respect and to that degree, technologies emerge as focal entities that may receive the multiple attentions humans give the different forms of the other. For this reason, a third formalization may be employed to distinguish this set of relations:

$$I \rightarrow \text{technology-(-world)}$$

I have placed the parentheses thusly to indicate that in alterity relations there may be, but need not be, a relation through the technology to the world (although it might well be expected that the *usefulness* of any technology will necessarily entail just such a referentiality). The world, in this case, may remain context and background, and the technology may emerge as the foreground and focal quasi-other with which I momentarily engage.

This disengagement of the technology from its ordinary-use context is also what allows the technology to fall into the various disengaged engagements which constitute such activities as play, art, or sport.

A first phenomenological itinerary through direct and focal human-technology relations may now be considered complete. I have argued that the three sets of distinguishable relations occupy a continuum. At the one extreme lie those relations that approximate technologies to a quasi-me (embodiment relations). Those technologies that I can so take into my experience that through their semi-transparency they allow the world to be made immediate thus enter into the existential relation which constitutes my self. At the other extreme of the continuum lie alterity relations in which the technology becomes quasi-other, or technology "as" other *to* which I relate. Between lies the relation with technologies that both mediate and yet also fulfill my perceptual and bodily relation with technologies, hermeneutic relations. The variants may be formalized thus:

Human-technology-World Relations
Variant 1, Embodiment Relations
(Human-technology) $\rightarrow$ World
Variant 2, Hermeneutic Relations
Human $\rightarrow$ (technology-World)
Variant 3, Alterity Relations
Human $\rightarrow$ technology-(-World)

Although I have characterized the three types of human-technology relations as belonging to a continuum, there is also a sense in which the elements within each type of relation are differently distributed.

There is a *ratio* between the objectness of the technology and its transparency in use. At the extreme height of embodiment, a background presence of the technology may still be detected. Similarly but with a different ratio, once the technology has emerged as a quasi-other, its alterity remains within the domain of human invention through which the world is reached. Within all the types of relations, technology remains artifactual, but it is also its very artifactual formation which allows the transformations affecting the earth and ourselves.

All the relations examined heretofore have also been focal ones. That is, each of the forms of action that occur through these relations have been marked by an implicated self-awareness. The engagements through, with, and to technologies stand within the very core of praxis. Such an emphasis, while necessary, does not exhaust the role of technologies nor the experiences of them. If focal activities are central and foreground, there are also fringe and background phenomena that are no more neutral than those of the foreground. It is for that reason that one final foray in this phenomenology of technics must be undertaken. That foray must be an examination of technologies in the background and at the horizons of human-technology relations.

D. BACKGROUND RELATIONS

With background relations, this phenomenological survey turns from attending to technologies in a foreground to those which remain in the background or become a kind of near-technological environment itself. Of course, there are discarded or no-longer-used technologies, which in an extreme sense occupy a background position in human experience—junk. Of these, some may be recuperated into non-use but focal contexts such as in technology museums or in the transformation into junk art. But the analysis here points to specifically functioning technologies which ordinarily occupy background or field positions.

First, let us attend to certain individual technologies designed to function in the background—automatic and semiautomatic machines, which are so pervasive today—as good candidates for this analysis. In the mundane context of the home, lighting, heating, and cooling systems, and the plethora of semiautomatic appliances are good examples. In each case, there is some necessity for an instant of deistic intrusion to program or set the machinery into motion or to its task. I set the thermostat; then, if the machinery is high-tech, the heating/cooling system will operate independently of ongoing action. It may employ time-temperature changes, external sensors to adjust to changing weather, and other cybernetic operations. (While this may function well in the home situation, I remain amused at the still-primitive state of the art in the academic complex I occupy. It takes about two days for the system to adjust to the sudden fall and spring weather changes, thus making offices which actually have opening windows—a rarity—

highly desirable.) Once operating, the technology functions as a barely detectable background presence; for example, in the form of background noise, as when the heating kicks in. But in operation, the technology does not call for focal attention.

Note two things about this human-technology relation: First, the machine activity in the role of background presence is not displaying either what I have termed a transparency or an opacity. The "withdrawal" of this technological function is phenomenologically distinct as a kind of "absence." The technology is, as it were, "to the side." Yet as a present absence, it nevertheless becomes part of the experienced field of the inhabitant, a piece of the immediate environment.

Somewhat higher on the scale of semiautomatic technologies are task-oriented appliances that call for explicit and repeated deistic interventions. The washing machine, dryer, microwave, toaster, etc., all call for repeated programming and then for dealing with the processed product (wash, food, etc.). Yet like the more automated systems, the semiautomatic machine remains in the background while functioning.

In both systems and appliances, however, one also may detect clues to the ways in which background relations texture the immediate environment. In the electric home, there is virtually a constant hum of one sort or the other, which is part of the technological texture. Ordinarily, this "white noise" may go unnoticed, although I am always reassured that it remains part of fringe awareness, as when guests visit my mountain home in Vermont. The inevitable comment is about the silence of the woods. At once, the absence of background hum becomes noticeable.

Technological texturing is, of course, much deeper than the layer of background noise which signals its absent presence. Before turning to further implications, one temptation which could occur through the too-narrow selection of contemporary examples must be avoided. It might be thought that only, or predominantly, the high-technology contemporary world uses and experiences technologies as backgrounds. That is not the case, even with respect to automated or semiautomatic technologies.

The scarecrow is an ancient "automated" device. Its mimicry of a human, with clothes flapping in the breeze, is a specifically designed automatic crow scarer, made to operate in the absence of humans. Similarly, in ancient Japan there were automated deer scarers, made of bamboo tubes, pivoted on a pin and placed so that a waterfall or running stream would slowly fill the tube. When it is full enough, the device would trip and its other end strike a sounding board or drum, the noise of which would frighten away any marauding deer. We have already noted the role automation plays in religious rituals (prayer wheels and worship representations thought to function continuously).

Interpreted technologically, there are even some humorous examples of "automation" to be found in ancient religious praxes. The

Hindu prayer windmill "automatically" sends its prayers when the wind blows; and in the ancient Sumerian temples there were idols with large eyes at the altars (the gods), and in front of them were smaller, large-eyed human statues representing worshipers. Here was an ancient version of an "automated" worship. (Its contemporary counterpart would be the joke in which the professor leaves his or her lecture on a tape recorder for the class—which students could also "automatically" hear, by leaving their own cassettes to tape the master recording.)

While we do not often conceptualize such ancient devices in this way, part of the purpose of an existential analysis is precisely to take account of the identity of function and of the "ancientness" of all such existential relations. This is in no way to deny the differences of context or the degree of complexity pertaining to the contemporary, as compared to the ancient, versions of automation.

Another form of background relation is associated with various modalities of the technologies that serve to insulate humans from an external environment. Clothing is a borderline case. Clothing clearly insulates our bodies from temperature, wind, and other external weather phenomena that could become dangerous to life; but clothing experienced is borderline with embodiment relations, for we do feel the external environment through clothing, albeit in a particularly damped-down mode. Clothing is not designed, in most cases, to be "transparent" in the way the previous instrument examples were but rather to have a certain opacity without restricting movement. Yet clothing is part of a fringe awareness in most of our daily activities (I am obviously not addressing fashion aspects of clothing here).

A better example of a background relation is a shelter technology. Although shelters may be found (caves) and thus enter untransformed into human praxis, most are constructed, as are most technological artifacts; but once constructed and however designed to insulate or account for external weather, they become a more field-like background phenomenon. Here again, human cultures display an amazing continuum from minimalist to maximalist strategies with respect to this version of a near-background.

Many traditional cultures, particularly in Southern Hemisphere areas, practice an essentially open shelter technology, perhaps with primarily a roof to keep off rain and sun. Such peoples frequently find distasteful such items as windows and, particularly, glassed windows. They do not wish to be too isolated or insulated from the elements. At the other extreme is the maximalist strategy, which most extremely wishes to totalize shelter technology into a virtual life-support system, autonomous and enclosed. I shall call this a technological cocoon.

A contemporary example of a near-cocoon is the nuclear submarine. Its crew lives inside, and the vessel is designed to remain at sea for prolonged periods, even underwater for long stretches of time.

There are sophisticated recycling systems for waste, water, and air. Contact with the outside, obviously important in this case, is primarily through monitoring equally sophisticated hermeneutic devices (sonar, low-frequency radio, etc.). All ordinary duties take place in the cocoon-like interior. A multibillion-dollar projection to a greater degree of cocoonhood is the long-term space station now under debate.

Part of the very purpose of the space station is to experiment with creating a mini-environment, or artificial "earth," which would be totally technologically mediated. Yet contemporary high-tech suburban homes show similar features. Fully automated for temperature and humidity, tight air structures, some with glass that adjusts to glare, all such homes lie on the same trajectory of self-containment. But while these illustrations are uniquely high-technology textured, there remain, as before, differently contexted but similar examples from the past.

Totally enclosed spaces have frequently been associated with ritual and religious praxis. The Kiva of past southwestern native American cultures was dug deep into the ground, windowless and virtually sealed. It was the site for important initiatory and secret societies, which gathered into such ancient cocoons for their own purposes. The enclosure bespeaks different kinds of totalization.

What is common to the entire range of examples pointed to here is the position occupied by such technology, background position, the position of an absent presence as a part of or a total field of immediate technology.

In each of the examples, the background role is a field one, not usually occupying focal attention but nevertheless conditioning the context in which the inhabitant lives. There are, of course, great differences to be detailed in terms of the types of contexts which such background technologies play. Breakdown, again, can play a significant indexical role in pointing out such differences.

The involvement implications of contemporary, high-technology society are very complex and often so interlocked as to fall into major disruption when background technology fails. In 1985 Long Island was swept by Hurricane Gloria with massive destruction of power lines. Most areas went without electricity for at least a week, and in some cases, two. Lighting had to be replaced by older technologies (lanterns, candles, kerosene lamps), supplies for which became short immediately. My own suspicion is that a look at birth statistics at the proper time after this radical change in evening habits will reveal the same glitch which actually did occur during the blackouts of earlier years in New York.

Similarly, with the failure of refrigeration, eating habits had to change temporarily. The example could be expanded quite indefinitely; a mass purchase of large generators by university buyers kept a Minnesota company in full production for several months after, to be prepared the "next time." In contrast, while the same effects on a shorter-term basis were experienced in the grid-wide blackouts of 1965, I was in Vermont at my summer home, which is lighted by kerosene lamps and even refrigerated with a kerosene refrigerator. I was simply unaware of the massive disruption until the Sunday *Times* arrived. Here is a difference between an older, loose-knit and a contemporary, tight-knit system.

Despite their position as field or background relations, technologies here display many of the same transformational characteristics found in the previous explicit focal relations. Different technologies texture environments differently. They exhibit unique forms of non-neutrality through the different ways in which they are interlinked with the human lifeworld. Background technologies, no less than focal ones, transform the gestalts of human experience and, precisely because they are absent presences, may exert more subtle indirect effects upon the way a world is experienced. There are also involvements both with wider circles of connection and amplification/reduction selectivities that may be discovered in the roles of background relations; and finally, the variety of minimalist to maximalist strategies remains as open to this dimension of human-technology relations as each of the others.

E. HORIZONAL PHENOMENA

The limits of this first analysis are now close at hand. With the indirect and field aspects of background relations, there are also hints of horizonal phenomena, which mark the boundaries of a phenomenology. (It should be noted that the term "horizon" in phenomenology is a limit concept. Unlike the common English expression "to expand one's horizons," no such possibility exists here. The horizon is the limit beyond which the inquiry ceases to display its internal characteristics, and like a voyage at sea, the horizon always remains that distant boundary between sky and sea or land and sea. It never comes nearer.)

Horizons, however, may be indicated in a number of ways with respect to extreme fringe phenomena. They are of important and indicative use here, such that mention must be made prior to moving to a second program. In the entire preceding analysis, the experience of technology has always been such that there is some experientially recoverable difference between what is experienced and the experiencer. All transparencies have been noted to be quasi-transparencies, all alterities quasi-otherness, and even the absent presences of background phenomena are at least indirectly recoverable.

Such a positioning of technologies as belonging to a continuum of relations, which points to their essential artifactual properties within human praxis, has been a constant found in all variations. Yet the question of the extremities beyond which there is no recovery, where perhaps technologies cease to be technologies, remains intriguing. I shall try to locate just such phenomena within the spectrum of the explicit phenomenologies I have followed.

In the case of embodiment relations, are there embodiments which fall "below" recovery? Embodiment technologies are taken into self-experience but remain distinctive, at least in fringe or echo phenomena, as other than my self. But what of the secret desire to have technology become myself? Such desires are approximated in the bionic trajectories of contemporary technologies. A borderline example is a crowned tooth. Although there is little doubt that something is being "added" to my mouth during the dental procedure and there is a period of accommodation in which I experience the strangeness of the cap, after a time the tooth becomes almost totally embodied. Yet even here, there remains a less-distinct recoverable awareness of the difference in texture, the feeling of hot and cold, and the slipperiness that calls for special chewing gums if it is not to stick. But the artifactuality of the cap is, indeed, an extreme to recover as a relation. It comes close to "being me."

Less so are the various implants which, users report, are experienced as a different mode of being one's own body. Hip joints of stainless steel and teflon do allow the previously restricted walker to resume close-to-normal walking, but the discernible difference remains a fringe awareness of the bionically transformed individual. Yet such prosthetic implantations do allow a being-towards-the-world as a modified bodily being.

Such mechanical transplants do not yet reach the extremity of horizonal phenomena. An even more extreme set of examples arise from chemical transformations, or what I shall call edible technologies. The history of the birth-control pill is instructive in this case. Early users of the pill reported two results: They did experience bodily changes, in that period pains were experienced differently; and sometimes there were other side effects such as minor nausea. But as with the previously noted fascination with the amplifying transformations of all new technologies, most such side effects were repressed in favor of the exultation over a worry-free ability to engage in close-to-"natural" or pregnancy-free sexual relations.

Later, delayed side effects were associated with the pill (in some cases, elevated blood pressure, etc.; later, worries over cancers), but these could at best be indirectly experienced. The pill, once taken, functioned as a kind of internal background relation of the most extreme fringe type. As with all edible technologies, the "I am what I eat" phenomenon placed most effects at a distance or were delayed.

This very case, however, is instructive in a different way. Are edible technologies technologies at all? They are clearly technological products (of the chemistry industry) and are clearly technologically transformed entities; but insofar as they are absorbable into our very bodies, they, much more closely than our previous artifactual examples, are deeply embodied.

Even more ambiguously, the question of biological technology must be raised at this horizonal extreme. I refer here not to the popular issues revolving around whether animal life, through gene manipulation, can or should be patented but to the now literal reflexivity which biological technologies may have upon the transformation of human bodily life itself. A genetic transformation does display the same structural transformational features that the other human-technology relations display, but insofar as a genetic result becomes what we are, it dips below the level of at least the contrastive existential analysis taken here. This is not to say, however, that a hermeneutically enhanced analysis is precluded.

With biological technics, there is reached a new boundary between technology and life where the horizons of nature and artificiality are blurred. Yet while the ability to manipulate genetic material more precisely through identification of particular genes or strands is contemporary (a feat made possible by the instrumental ability to magnify the micro-levels of being, hence, conditioned by an essential possibility of technology) such biological technics are also ancient. Plant and animal breeding, hybridization, and other manipulations of wild products are as old as civilization and occurred independently in many parts of the world.

Furthermore, the blurring of the borderline life/artificial product is anticipated in the changes of attitude and use towards domestic plants and animals. They become clearly "use beings" under human control and for human use.

Horizons belong to the boundaries of the experienced environmental field. Like the "edges" of the visual field, they situate what is explicitly present, while as a phenomenon itself, horizons recede. And whether we refer to a kind of inner horizon (the fringes of embodiment) or the extremities of the external horizon (the ultimate form of texturing that a specific technological culture may take), the result is one of "atmosphere." This does not mean that the elements, particularly in terms of social phenomena, cannot be identified. For example, in contemporary situations, one atmospheric phenomenon is the vague fear or anxiety related to the possibility of nuclear threat. This fear is felt by children as well as parents and across at least all industrialized cultures. It is part of the texture of the technological culture that has an ultimate and universalizable destructive magnificational power. But this atmospheric social fear phenomenon is an effect of something else: the dream of technological totalization which marks one feature of maximalist technological cultures.

There are a number of ways in which nuclear fear might be historically analyzed. Obviously, destruction of one's enemies has been a human aim as ancient as humanity. Equally obviously, it has never been possible until the twentieth century to actualize this aim universally. Even Samson tires after thousands; but the potential, however quickly (full exchange of superpower weaponry) or slowly (nuclear winter), to eliminate the species is only possible for the present and the future. The inherent problem of a universal totalized solution is equally obvious—it is self-reflexive and applies to the killer as well as the killed. The political necessity to preserve such a universally genocidal means of "defense" takes the shape of attempts to deny self-reflexivity. Governments must find ways to convince populaces that there will be survivors (nuclear winter will be only autumn, etc.).

Yet the atmospheric nuclear fear is but the underside of a previous century's utopianism that a technological culture, once totalized, would be able to solve all basic social problems such as hunger, disease, a decent standard of living, and peaceful interchange between nations. If such hopes are reduced to single system utopianism in the late twentieth century, they nevertheless spring from the same mythical sources as the current fears.

Γ. EVE AND THE SPACESHIP

Perhaps what is needed at the end of this first itinerary is another fictive variation: to imagine the most extreme maximization of technology, for the other end of the continuum that began in the imagination of a non-technological Garden. This projection must aim at a totalized technological culture. The problems of even imagining such a state of existence are parallel to those found with the dream of technological innocence at the beginning, for we must project from our present situation in which there is only a presumption of totalization.

Yet there are clues to a trajectory of totalization from precisely this present. The imagination, science-fiction-like, would have to find technologically totalized components for each dimension of the existential situation of human beings in the world previously surveyed.

For World, we would have to have an artifactual "world," or self-contained environment, that was totally autonomous within some complex of technological conditions. Although we are no more able to point to such an artificial "world" than we are able to find any actual Garden, the concept of the technological cocoon is the instance of such a trajectory. It is even part of our fictional tradition and now a part of our economic debate (how many billions are we willing to spend on such an experiment?). At the present, such a situation would be temporarily lived in and clearly would be highly dependent upon earth stations, but the dream is for totalization. Yet approximations to fully enclosed mini-cocoons are, in fact, the recreational vehicles, the fully controlled building environments, and even certain kinds of condominiums we actually do inhabit. Insulated as much as possible from weather (unless it is pleasant) the projected artificial world is but an extension of present conditions. It is also a reflection of present desires.

At first, this somewhat ironic projection could be taken to hide some form of nature romanticism, a form which frequents even the widest space fantasies of many cinema and television series. Whereas the focal plot and visual effects are all of projected high-technology spaceships, alien beings, etc., the theme all too often is a quest for a lost earth or a new earth. Romanticism also belongs to utopian thought. But that is not the point; rather, I wish to point up that insofar as we are successful in actually constructing approximations to total- ized or at least maximized technological cultures, the result is neces- sarily to replace the dominant regions of threat and problem from "nature" to precisely technological "culture."

In the last few years, the triplet—"Challenger," Bhopal, and Chernobyl—has become almost symbolic of this replacement of threat phenomenon. Each breakdown was felt and taken up as a peculiar late-twentieth-century event. Bhopal and Chernobyl, by far the largest catastrophes in terms of human impact, became equivalents of the past major natural disasters such as hurricanes or tidal waves, sweep- ing the world with their aftereffects. Fears of breakdown on a large scale become fears of technologically textured societies.

These events, which occupy worldwide news, are indicative of a subtle but clear replacement or at least equivalence of past threats largely from natural disasters. The same mix of wider and instant com- munication, even to the interinvolvement of media and event, occurs in the strange archaic/contemporary movements of much international terrorism. Revivals of religious fundamentalism, whose feuds now are spread with contemporary weaponry and publicized by contemporary media (from the tape-cassettes that spread Islamic fundamentalism un- derneath the controlled news media of the Shah, to the linkage of ter- rorist publicity, to the need for instant television coverage). This, too, is part of the substitution process.

Not all interlinkage, however, is symbolized by breakdown. Within the same tragedies were the internationalization of health care, with American bone marrow transplant experts helping Soviet patients and with rock stars stimulating a vast worldwide food transport pro- gram for Ethiopian starvation victims. The interlinking of media and event here is just as close as in the above negative examples.

If the technological cocoon is the microcosm for an artifical world in the dream of a totalized technological culture, its other components have also received imaginative treatment. If the cocoon is a replace- ment for a wider world, then the inhabitants must undergo similar ex- trapolative treatment.

When we turn to examples of the new technological Eves who in- habit our spaceship cocoons in science fiction—cinema or television form—they are frequently hybrids or imitations (androids or robots cleverly contrived, as Descartes imagined, to be mistaken for humans).

The bionic being, whether woman or man, is a being conceived of with new technological parts that unlike all extant such parts, have improved upon the weakness of the replaced part. Such imaginings of hybrids, more powerful and more perfect than the previous merely biological beings, are reflections of precisely the technological dreaming earlier noted. The bionic beings have become perfect unions of technology and life such that they simply are and experience themselves as superbeings. The technology has ceased to be a technology and become a (more perfect) human.

Fictional androids and robots, however, are nearly perfect replicas or semblances of humans, although most plots eventually reveal them to be Cartesian-clever machines, usually with some hidden flaw. These entities, which in this scheme would be even more alterities than actual technofacts, begin to reveal the underside of fear that is directed at the aim of the dream itself. What if we were able to actually equal or exceed ourselves through our own inventions? Would our inventions then find us of no use or, worse, of use to them?

Take the imagination of technological perfecting in a different direction. What if we were technologically able to duplicate a human being, including all the foibles and warts? Would we not be in the perfect reading-machine situation? What sense would there be to a technology perfectly duplicative of something non-technological? At best, such an invention would satisfy the copy urges of the designer, as a perfect forgery would the forger.

But if the android were simultaneously both human and transformed, then it might be of interest. And here the fictional varieties abound: the perfect human slave, or the perfect human lover, or the perfect human strong person, etc. Yet here again we reach the contradiction of the dream in which the desire is both to have and not to have a technology, with its artifactual and selective essence. No human Eve would ever wish to be a "Stepford Wife."

Although the above fictive variations have been directed at the others who might cohabit within the cocoon, what has been revealed there applies in a similar way to a possible technological totalization of the self. Which Eve in the spaceship would voluntarily yield to a total technological transformation? The answer is, in contemporary life, somewhat ambiguous. The use of steroids by athletes is apparently widespread in spite of known risks (later muscle deterioration, high blood pressure, increased risk of heart failure). The use of cosmetic surgery, including implants, is fashionable. On a much more ancient level, the human uses of everything from makeup to initiatory masks and scarification are instrumentally applied transformations of the human body. Yet none of these practices comes close to the totalized variation, for beyond the above probably trivial pursuits, few Adams or Eves would willingly submit to heart, liver, or prosthetic transplants except when the loss of life or motility threatens.

That we should stop short of totalization is not itself strange; what is strange is the persistence of the modern Cartesian dream of the perfectability of the cleverly designed automaton as a substitute for the human being.

G. DREAMS OF TOTALIZATION

Within this first itinerary through human-technology relations there appears to be a vast anomaly. On the one side, an examination of human-technology relations reveals a deep continuity in the ways humans have experienced their technologies. These existential relations apply equally to ancient and modern, to simple and complex technologies. Insofar as a structural phenomenology aims at invariants, this is as it should be; yet this emphasis upon sameness leaves something out—particularly regarding what appears to contemporaries as the vast difference between all traditional and nonindustrial societies and our own.

In part, this impression is part of a tactic of this first program in human-technology analysis. I have tried to show that there *is* a continuity that spans both human history and cultures beyond the extravagant claims of those who would dissociate us from our pasts and our peers. Yet to disregard the disjuncture between today's maximalist technological cultures and those of minimalist or less-than-minimalist lifeforms would also be distorting.

Not that hints are lacking regarding these differences. One such theme, followed somewhat narrowly but purposely here, has been the emphasis upon the disjunctions made possible through modern instrumentation. Modern science, in my view, is distinctively different from all ancient science by virtue of its embodiment in instrumentation, whether of embodiment or hermeneutic types, for these have made possible views of the universe that both imaginatively and perceptually exceed and differ from all ancient cosmologies. Alfred North Whitehead has made the same point:

> The reason we are on a higher imaginative level is not because we have a finer imagination but because we have better instruments. In science, the most important thing that has happened in the last forty years is the advance in instrumental design . . . a fresh instrument serves the same purpose as foreign travel; it shows things in unusual combinations. The gain is more than a mere addition; it is a transformation.[12]

That transformation is a quality disjunction that begins to occur with the rise of modern science but is greatly accelerated only in more recent times.

A second clue to a disjunction between ancient and contempo-

---

12. Alfred North Whitehead, *Science and the Modern World* (New York: New American Library, 1963), p. 107.

rary times lies in the ways dreams of totalization are actualized in today's maximalist and yesterday's minimalist cultures. Indeed, the fact
that humans have become a virtual geological force through the magnified technologies of recent time is the seat of one of our distinctive
contemporary worries. If industrial products are a major cause of the
newly discovered ozone hole in the Antarctic as suspected, if the
threatened elimination of a number of small animal species because of
rain forest destruction (a percentage roughly equivalent in number to
the past's large animal species extinction), if acid rain creating pollution threatens most higher-altitude northern hemisphere forests, then
we have reason for this worry.

Yet while the intensity of the human-through-technology impact
in today's world is clearly of a magnitude greater than in any historical
time, it is not without precedent. J. Donald Hughes, in a remarkable if
pessimistic book, *Ecology in Ancient Civilizations*, argued that regard-
less of cultural differences, each of the civilizations surrounding the
Mediterranean produced the same results of (a) deforestation, (b) over-
grazing by sheep and goats, and (c) irreversible erosion resulting in
(d) today's arid climate all around the Mediterranean basin.

This result was similar whether we speak of Greeks, Hebrews, Romans, Phoenicians, or any of the other progenitors of our later European civilization. Ecologically speaking, Hughes provides an anti-
romantic counterpart to Heidegger's glorification of the Greek temple:

> Those who look at the Parthenon, that incomparable symbol of the
> achievements of an ancient civilization, often do not see its wider setting.
> Behind the Acropolis, the bare, dry mountains of Attica show their rocky
> bones against the blue Mediterranean sky, and the ruin of the finest
> temple built by the ancient Greeks is surrounded by the far vaster ruins of
> an environment which they desolated at the same time.[13]

Even more ancient are the implications of humans in the extinctions of the large mammals of the late Ice Age. Or, to turn to a more
recent and non-Western example, similar extinctions of all the large
wingless birds on most Pacific islands occurred with the westward expansion of the Polynesian navigators of whom I have earlier spoken.
Such results are clearly not the unique effects of either our cultural ancestors or our histories but lie deeper in the possibility structures of
the human.

Do such trajectories arise from the dreams of totalization suggested here? Again, the answer must respect cultural ambiguity. To
complete the first itinerary, I wish to contrast what I shall call two different dreams of totalization. I shall take as my examples our own
technologically maximalist one and contrast it with one of the most

---

13. J. Donald Hughes, *Ecology in Ancient Civilizations* (Albuquerque: University Press
of New Mexico, 1975), p. 1.

technologically minimalist cultures known to me, that of the inland
Australian Aborigines.

For illustrative purposes only—because I find deep flaws in such
an interpretation—I shall assume that those critics of our historical development are correct in characterizing the dominant trend of our
Technological (with the reified "T") culture as one whose dream of totalization is geared to bringing nature itself under the control of culture. If the purpose of technology is to control nature or to make it
available to human purposes, then interpretations such as those of
Lynn White, Jr., who saw in Western history a search for power arising
out of religious consciousness (to subdue and dominate the earth), or
that of Heidegger, who saw the revealing power of modern technology to have revealed the world as a "resource well" (my translation of
Bestand), are correct.

If one then takes the inland Aboriginal cultures as a contrast, it
might seem that here lies an exception to precisely the direction taken
by our histories. So taken, the example is impressive: First, one must
note that the ancestors of the Aboriginals arrived at least 40,000 years
ago, as evidenced by artifacts found in the south of Australia. Second,
they were largely isolated from external invasions and migrations (an
exception apparently occurred in the north some three thousand years
ago when the dingo was introduced; this non-neutral introduction of a
wild dog led to the extinction of the marsupial wolf, one of the few
marsupial predators). Thus, the Aboriginals may be said to be one of
the world's longest-lasting isolated but continuous cultures.

That they survived at all in such arid conditions is impressive, but
to have survived in such a way that their primary culture could be
characterized by what we might call "leisure activities" is even more
noteworthy. In that context, it should be noted that very few hours of
the day are spent gathering or preparing food. Much more time, relatively, is spent in their version of philosophical argument—times of
storytelling and arguing over a religio-mythical system noted for its
complexity. Celebratory activities take up much of the communal cultural time. Weddings, funerals, seasonal festivals were noted to last
weeks on end at the time of the first European explorers. Artistic activity, too, was highlighted in both visual and auditory form. Arnheim
Land's "X-ray" paintings and didjereedoo music are today valued by
denizens of our postmodern culture.

Yet this long-lasting culture was clearly one of the most minimalist
of technological cultures. Not that it lacked material artifacts al-
together: There was a weapons collection, including the invention of
both returning and non-returning boomerangs, spears (pre–Stone Age;
tips were hardened wood), spear-throwing devices to amplify arm
power, slings, etc., but not the bow and arrow found virtually worldwide in other ancient cultures.

And if the men's repertory of hunting weapons is to be matched by an equal set of technologies on the female-gathering side (and this more important food source), then the advanced basketry, adapted emu egg carriers, digging, and other tools are also to be noted.

Within religious practice there were thuringas and masks and a wide variety of aesthetic-religious artifacts, including the famous bark paintings and cave drawings of "X-ray" animal depictions. Yet the spectrum is minimal by any comparison to any other non-migratory peoples' set of tools or manufactured objects.

"Technical" knowledge, however, was vast and complex—particularly with respect to food sources. As I began to study this culture, I was tempted to speculate that one central cultural axiom may well have been: "Everything is potentially food; the only question is how to make it so." The Aboriginals discovered or invented detoxification processes that sometimes took three steps, in contrast to many other cultures' single heat-by-boiling-or-baking or soaking techniques. The range of food sources included some thirty varieties of breads made from grass and weed seeds, delicacies of baked bats caught by children, and the hard-to-find witchetty grub (a five-inch-long grub inhabiting the root system of the witchetty tree). The grub is found by stomping the heel of one's foot around the base of the tree and listening to the echo-located hollow sounds.

Tales are told of the amazement of the Aboriginals at the first Europeans to try to cross the central deserts of Australia. Many of those folk died of thirst, desiccation, and exposure. The Aboriginal response was wonder at how, in the land of such plenty, anyone could be so stupid. For water, for example, all one had to do was to find the crack in the dried mud floor of a proper location and dig down some two feet to a membrane-enclosed, water-surrounded frog awaiting the next rains (which might be several years apart), drink the water, and eat the frog for a chaser!

Here, then, is a minimalist culture that of necessity learned to cope with a harsh environment. For the culture to succeed, however, there were prices to be paid. A second axiom, after maximizing the edible, must be one directed at minimizing population growth. Here, as expected, are means to be found: An unusual one is a form of "male birth control" in the practice of ritual subincision of adolescent male children in puberty rites. Subincision consists of a permanent hole made at the bottom base of the penis through the use of a thorn inserted during the initiation ceremony. Thereafter, a true "man" will have to hold this hole when urinating to mark him as a man. An unexpected side effect is that there will be a limited amount of sperm escaping during intercourse, thus effectively lowering the sperm count and hence lowering the probability of insemination. Of course, more drastic measures also had to be taken to keep the population down. Associated with the belief that no couple should have more than two children (who could be carried in flight; a third would theoretically have to be left behind), there is evidence of another familiar South Pacific means of birth control: infanticide.

Such prices are unlikely to be paid by those of us inhabiting our own cultures. Similarly, life expectancies, tolerance of weather extremes, and the like are not about to be traded by us for survival in such a lifeform. And the ultimate intercultural answer also seems visible among Aboriginals today as well—it is a culture largely subsumed within the fringes of a now two-century-old European dominant culture.

I shall not disguise a certain admiration for the human learning and adaptability that inland Aboriginals exemplified, particularly in relation to what I would call a religio-ethical system that functioned as a conservationist ethic. Two simple beliefs indicate how such an ethic functioned. One rule relates to the prohibition against killing any animal outside its sacred territory. The context for this rule is explained in terms not likely to be believed by us. It includes elements of totemism, in which the animals noted belong to a kin group (although the close relationship between humans and animals is a matter coming into more interesting question in our own time).

The way this conservation rule functions is clear. Animals can be eaten—but only within limits. As it turns out, a sacred territory is precisely the animal's breeding and water-hole territory. In times of severe drought or in breeding seasons, the animal must be spared, even in extreme conditions, to the point that even some of the tribe must be sacrificed for the sake of the animal. In this way a balance of human/animal populations is insured, in turn insuring a future.

A second, similar rule, again grounded in totemistic beliefs, relates to the rare unused or unusable animal. Leeches, for example, are not to be unnecessarily killed. And the reason given is that leeches belong to the great series of relationships that characterize the Aboriginal world: they are part of a close-linked ecosystem. Such an understanding has deep functional significance in a delicate desert environment. It is cast in prescientific terms but is "contemporary" with respect to sound ecological understanding.[14]

The point here, however, is not to recommend substituting Aboriginal culture for ours nor to simply romanticize the minimalist mode of life exemplified by that culture; rather, it is to make a contrast with our culture with respect to the relative role played by technologies. It is not simply the absence of complex technologies that makes the two cultures different but what could be called the two radically different

---

14. A more complete discussion of the specifics of Aboriginal ethics may be found in David H. Bennett's "Inter-Species Ethics: A Brief Aboriginal and Non-Aboriginal Comparison," Discussion Papers in Environmental Philosophy, No. 7, Australian National University, 1985.

ways in which the world is revealed. The Aboriginal, too, follows a
dream of totalization, but a different dream—that of the Dreamtime.

The Dreamtime is "nature" brought into "culture" in a way different from the Heideggerian understanding of a resource well. Yet all of
nature is "taken into account" in the Dreamtime, through stories that
make the features of the landscape into sacred sites. Thus, the promontory is "where Goanna dreams" in a long and complex tale that is
told and that also exemplifies some moral and tribal value. The water
paths are "where the great snake wandered" in some other set of similar tales. The Dreamtime is a nontechnological taking of the world of
nature into this culture, yet it is an attempted totalization, a mode of
world revealed.

Functionally, such a mode of revealing a world, particularly when
considered over the 40,000-plus-year history of these peoples, did little to modify the actual environment. And it was a cultural history
which did not develop the fascination for artifactual invention shown
in our own history.

So, at the end of this first itinerary, at the least, there is both a
sameness to our trajectories in dreaming of totality and a deep difference. The dream of taking nature into culture technologically is shown
to belong to a history, to *our* history. The enigma that appears at this
junction is that our very distinctive history seems a currently dominant
one. Aboriginal culture today is vestigial in the sense that, with isolated
exceptions, the old ways are no longer being transmitted, at least not
in their original forms. Even the vibrant religio-art of the peoples must
now fit into a context which re-situates it into a crosscultural exchange, to be considered as an art commodity no longer belonging to
its dreamtime relationship. But that is the fate of virtually all traditional
cultures of this type.

Is there, then, a single, massive trajectory to the rise of high-technological culture and its attainment as a world culture? If so, the
Marcuses, Jonases, and Elluls would turn out to be the prophets for
our times.

6. Program Two: Cultural Hermeneutics

The first program in this analysis followed a more strictly phenomenological emphasis upon the human experience of technologies.
With the second program—a cultural hermeneutics—a shift of perspective is called for. Here the question revolves around the ways in
which cultures embed technologies. I have been using the term
"technology" in a broad sense, equivalent to certain aspects of material culture. From the outset, technology in the ensemble is seen
as belonging in an intimate way to our cultures. Nor have cultural
variations been lacking in the previous examples. At the least, I have
taken note of the spectrum from maximalist to minimalist technology-culture embeddings.

At the cultural level, however, more occurs than simply the number and type of human-technology relations. I have contended that insofar as there are a limited number of types of human-technology
relations following from human existential structures, all cultures exem-
plify the full range of these relations (invariantly), although the mixes
are clearly highly variant. Here the focus will be on non-consonant
variations, seeking the patterns that inhere in cultural diversity.

Again the history and anthropology of technology will be used—
as well as imaginative variants. And my aim, as in the first program, is
to suggest a different framework of interpretation, one that can give a
new perspective to certain of the contemporary questions now being
directed at technological culture. The most general such set of questions revolves around the notion of an overall trajectory to technological history: Is the coming to dominance of Western-originated science
and technology a "fate" for the entire earth? And is such as trajectory
single? Related to these questions are those about the "autonomy" of
technology. Can we "control" technology? Or does it control us? Or
has technology become Technology with its own autonomy—runaway
like a Frankenstein?

In response to each of these current concerns about technology, I
propose a shift of perspective and, with it, a rephrasing of the questions. Such is the task of a cultural hermeneutics. I shall continue this
inquiry as into technology and the lifeworld, now with a primary shift
of focus to the macroperceptual field within which our bodily involve-

ments take place. The first inquiry looks at the phenomenon of technology transfer.

A. TECHNOLOGY TRANSFER:
TECHNOLOGIES AS CULTURAL INSTRUMENTS

In 1930 the first Australian gold prospectors entered the New Guinean highlands in search of gold. They were totally unaware that those highlands were peopled by nearly a million other humans organized into multiple tribes and who had never sighted "white" men before. The Australians came equipped with a movie camera and recorded some of these first encounters (later to be incorporated into a television documentary, which added interviews with the still-living persons who could recall these events).

For the New Guineans whose world had been thought to be the only world, the sight of these pale figures was astonishing—and they interpreted them to be either spirits returned from the dead or deities from some "other" world. (This response pattern has been noted in a number of both ancient and more recent encounters and often gives the invaders an initial advantage.) Only after following the newcomers and observing them for several weeks did they discover that they were, indeed, mortals. An Australian was spotted doing his morning excretion, and his droppings were discovered to have the proper "human" smell, thus concretizing a new understanding. Gods, after all, do not urinate, defecate, or fornicate!

Such droppings were not the only items curiously inspected by the tribesmen: every piece of refuse and artifact was inspected. Gifts of steel knives, in classical cross-cultural fashion, were enthusiastically received, as were steel axes. But, curiously, the first response to a rifle was ambiguous. In one of the film clips, a prospector demonstrated the power of a rifle. He shot a pig (close up, muzzle no more than inches from the animal's head); and while the shot's report caused some amazement, the ability to kill a pig at such range did not apparently seem impressive. Only later, after the prospectors were seen to be humans, were their artifacts perceived as desirable and worth a raid. Such a raid was organized with disastrous but predictable results: several New Guineans were killed by rifle fire. After this event, rifles were perceived as powerful weapons, worthy of having.

If the immediate acceptance of steel knives and axes is understandable and if the gun example remains partially ambiguous, more curious was the response to the oval sardine cans left behind by the Australians. These were immediately snatched by the New Guineans as treasured objects—and promptly made into the centerpieces of the elaborate headwear they wore for special occasions. The sardine can was placed at the center of the forehead piece, a position formerly occupied by the large shells of similar shape.

A standard analysis, adumbrated by a few Heideggerian insights, is easily sufficient to account for the steel axe and knife examples. These tools were immediately perceived by their shapes and functions to fit into an extant praxis. This is not a bare perception à la Heidegger's hammer, which must belong to a context of involvements, but is a perception sedimented within a known praxis. What the standard economic and cultural analysis will add is that by receiving the steel axe and knife, the New Guinean is receiving more than a mere artifact. He is potentially receiving a set of cultural relations—in this case, relations that will create a situation of dependency upon the incoming culture. ("Beware Australians bearing gifts"?)

Just as there is no such thing as "an" equipment, neither is there an equipment without its belonging to some set of culturally constituted values and processes. As the steel axe and knife become more desirable and since their mode of production is not available to the New Guinean, the new objects must be attained by some form of exchange, the histories of which are multiple enough not to recite here. But in addition to now tying into a new set of cross-cultural exchanges, one must also note that the context of the previously familiar object—in this case, stone knives and axes—also changes value and position. Whatever ritual aspects and beliefs surrounded stone tool making are fated to disappear with the unused equipment.

A similar borrowing by the Puluwateans, South Pacific navigators, better illustrates the culture-perceptual change accompanying such technology transfers. The Puluwateans steered by wave patterns, without a compass. Once becoming acquainted with the compass, these navigators adopted its use—largely because the compass was at first an object of fascination rather than something useful. A compass conferred prestige. But once it had been adopted and used for one of its purposes—to steer a straight course—it became possible to unlearn (de-skill) the more difficult wave perceptions, which were part of a complex initiation process in seamanship. The complex network of involvements to which the compass belongs in Western navigation, that is, the mathematical referents to the complex of directions (polarly focused), latitude and longitude, etc., was never adopted at all by the Puluwateans. The compass thus has a different "being" in Puluwatean use than in Western use.

Returning to the New Guineans, the remaining two examples become understandable within the first approximations to extant praxis as well, as long as we do not understand praxis to be modeled solely upon economic and efficiency considerations. The sardine-can-become-headwear example embeds a new artifact into an extant "fashion" praxis. Such a praxis incorporates into its context of significance the status and identity of the wearer in ways not necessarily too far from our own fashion intentionalities.

The rifle example also becomes understandable, although in a slightly more difficult way, if it is understood to be related to an already familiar praxis. Warfare and the killing of enemies were well enough known to the New Guineans, but the form and function of the rifle were not familiar as such. Only when it became apparent that this new weapon could be used at a distance (somewhat like a spear) could its value be seen. The function, at least, became perceivable once its use-context was fully demonstrated.

The adaptation of a transferred technology—at least, at first—depends upon its being able to fit into an extant praxis. But even when it is adapted, the context of significations may differ quite radically relative to the sedimented type of praxis in the recipient culture. One does not need to go to the more exotic cultural examples to take note of this phenomenon.

The works of Lynn White, Jr., Joseph Needham, and Daniel Boorstein have amply documented historical technology transfers that show such cross-cultural disjunctions in our own history. In his "Cultural Climates and Technological Advance in the Middle Ages," White points to what he discerns as the hunger for power and practical application in the Latin West, contrasting with an often more meditative context in both the Greek and the Far East. This transfer occurs in the new uses that borrowings were put to when transported to the West.

The previously cited example of technologically transferred Indian prayer wheels becoming windmills in the West is also an example of two different cultural embeddings. In the West—already anticipating seeing nature as a resource well, long before so credited by Heidegger—wind power was adapted to a variety of power uses such as pumping water, grinding grain, sawing lumber. White points out that Islamic countries—Iran and Afghanistan—also invented a windmill, but in spite of the obvious need of power for irrigation, these power sources were never transferred from the local sources of use elsewhere in Islamic lands and thus did not become what they were in Europe.[1]

White similarly traces the history and development of what were to become the flying buttresses of the great Gothic cathedrals from the ogival arch (India, second century) through a gradual migration westward to a church in Monte Casino (1071), to its adaptation to the unheard-of heights demanded at Cluny (1088). Yet what was to become the peak of cathedral building also owes its past to a borrowing, replaced in a different expressivity in Europe.[2]

The cultural desire for expressed power is clearly stated in the order for the building of the Duomo in Florence: "The Florentine Republic, soaring ever above the conception of the most competent judges, desires that an edifice shall be constructed so magnificent in its

---

1. Lynn White, Jr., "Cultural Climates," p. 176.
2. Ibid., pp. 183–85.

height and beauty that it shall surpass anything of its kind produced in the times of their greatest power by the Greeks and the Romans."[3]

In each of these examples, a "technology" has been transferred, yet there is also a sense—insofar as there is a doubled set of contextual involvements—that "a" technology is not what is transferred. The doubled context is this: First, there is the involvement of the artifact within its immediate use-context. It "is" what it is in relation to that context. Insofar as such contexts, particularly at the simplest levels, may be widespread with respect to cultures, a transfer is a relatively simple matter (steel axes and knives for stone ones). It is less so if the context of instrumental involvements are themselves complex, implying particularly some specific learned hermeneutic process (the compass example). In each case, the artifact becomes technologically what it "is" in relation to the degree and type of transferability to which the respective cultures overlap in practice. Here the overlap may be minimal.

But second, there is also the juxtaposition of the larger cultural contexts, which may not at all overlap. And the artifact "is" what it is also in relation to this cultural field. There is thus a sense in which the "windmill" as windmill was not transferred. The mere technical aspects of the prayer wheel do not become a windmill until reconstituted within the new cultural context. This contrast has been noted by Joseph Needham's studies of a developed set of technologies in China as compared to Europe: gunpowder and rocketry for celebration contrast with the same materials fitted into siege and warfare.

The temptation may be strong here to leap to a contextless conclusion that the "technology" as such is "neutral" but takes on its significance dependent upon different "uses." But such a conclusion remains at most a kind of disembodied abstraction. The technology is only what it is in some use-context. Even discarded technologies, whether in museums of science and industry, ruins against a landscape, or re-fitted into a bricolage construction, continue to indicate their perceivable "usefulness."

In this contrast of technologies transferred between cultures, clues begin to emerge regarding higher levels of technological complexity. Whereas the previous phenomenology could, and properly does, reduce its analysis to human-technology experiences (experiences in which both simple and complex technologies may fall into simple embodiment or often easily learned hermeneutic relations) in the phenomenon of technology transfer, the more complex the technologies, the more difficult to effect a transfer. I am not here implying that complex technofacts are not introduceable. To the contrary, precisely because they are often easily introduceable, the deeper problems of technology transfer emerge.

---

3. Mary McCarthy, *The Stones of Florence* (New York: Harcourt, Brace, Jovanovich, Publishers, 1959), p. 68.

If the easily transferable objects in past years of exploration could be steel knives and axes augmented by hawksbells and beads, mirrors, and the like, in today's neocolonialism the objects of fascination are more likely to be wristwatches, radios and televisions, calculators and computers. These transfers repeat in their own more contemporary fashion precisely some of the characteristics seen before.

The radio and the watch carry particularly fascinating features. The watch, even in a nonclock and loose timekeeping context, becomes an admired bauble. Its motions and sounds are fascinating; as a bracelet, it is also a prestige fashion object. The fashion praxis into which a watch fits is virtually universal; the radio even more so. It can convey sounds already familiar to the recipient. I shall never forget the first time I heard a Xhosa preacher intoning on African radio.

The new artifacts, while implying complex infrastructures of high-technology culture for production, do not yet at this level transcend the simple use-object transfer of praxis. Thus there can be a flow of artifacts and technofacts that do not yet massively transform the cultures. In water transportation, for example, while sail has often been replaced by motor power—and the varieties of power are highly variant, from outboards to the Southeast Asian "longtails," which use auto engines for high-speed performance—the types of crafts that are powered are still the native craft of the region; and the loads and cargo are little changed since earlier times. So long as this still characterizes the practice, those who argue for "appropriate technologies" may have a point. The cultural interface, however, takes place at two levels: the level of instrumental involvement, which we see has many overlaps at daily levels, and the more complex level of higher cultural values and their attendant complexes. It may make little immediate difference if a wristwatch is worn as a fashion object, but if it successfully carries in its wake the transformation of a whole society into a clock-watching society with its attendant social time, then a large issue is involved.

What is needed here is a more complex example in which the "same" technologies are embedded differently in roughly equivalent developments of culture. A brilliant example of this relates once again to that primordial invention, the clock, and the way in which time is perceived and used. We have already noted the Western example via Mumford and Heidegger and the way the clock became an instrument for regulating social time and an account of nature itself. Daniel Boorstein has provided a different example, the way clocks were embedded in Chinese history prior to the modern era.

Ancient Chinese civilization was, in its context, an amazingly sophisticated one and, not unlike most ancient civilizations, highly developed in its knowledge of astronomy or "heavenly" phenomena. As in many other ancient civilizations, the heavenly movements bespoke much more than mechanical movements of the stars and planets. A near universal for ancient astronomy is some version of astrology in which heavenly movements are thought (hermeneutically) related to existential processes. Chinese astronomy was accurate enough that the oldest-known prediction of an eclipse was recorded in 1361 B.C. The hermeneutic instrument which records and reflects these movements is the calendar; and, as we shall see, it is an accurate calendar, which in turn relates to the first Chinese clock.

That clock, purportedly invented by the Imperial Astrologer Su Sung sometime shortly after 1077 (not too different a date from the invention of the first mechanical clocks in Europe), was not for telling hours but for setting the astrological calendar for an Imperial need:

> The Emperor himself had an especially intimate need for calendrical timekeepers. For every night the Emperor in his bedchamber had to know the movements and positions of the constellations at every hour—in precisely the way Su Sung's Heavenly Clockwork made possible. In China the ages of individuals and their astrological destinies were calculated not from the hour of birth but from the hour of conception.
>
> When Su Sung constructed his imperial clock, the Emperor had as attendants a large number of wives and concubines of various ranks . . . including one empress, three consorts, nine spouses, twenty-seven concubines, and eighty-one assistant concubines. Their rotation of duty, as described in the Record of the Rites . . . was as follows:
>
> The lower-ranking come first, the higher-ranking come last. The assistant concubines . . . share the imperial couch nine nights in groups of nine. The concubines . . . are allotted three nights in groups of nine. The nine spouses and the three consorts are allowed one night to each group, and the empress alone, one night. On the fifteenth day of every month, the sequence is complete, after which it repeats in reverse order.[4]

A complex job for the timekeepers, clearly calling for a clock!

After this peak of clock development and use, however, the clock was abandoned in China and reintroduced only much later by the Jesuits at the time of Galileo. But again, while the clock became the machine for the Jesuits, its reintroduction remained tied to curiosity relating to automata (another near-universal curiosity feature between cultures). Soon a flow of cheap clocks from Europe to China began, where they remained largely *objects d'art*.

In short, the clock "is" what it is in relation to its embedded cultural matrix which, in China until modern times, remained very different from that of the West. A short glimpse at some of those contrary values may be instructive: (1) Official centralization in the office of the Emperor kept all calendars the property of the imperial house. (2) Emperors were kept hidden from the public and, while powerful as decree givers, isolated from public life. (3) Calendar-keeping was related

---

4. Daniel J. Boorstein, *The Discoverers* (New York: Vintage Books, 1985), p. 76. This book is a tour-de-force in the history of technological inventions and the role they have played, not only in Western but also in world cultures.

to the astrological features important as social predictions, particularly focused within the imperial confines. So long as the clock was kept to this context, it too was isolated.

Each of these factors stands in direct contrast to the Latin Western introduction of the clock which was public, timekeeping, and socially adapted. Here, then, is an example of a clearly important technology which "is" what it is within a cultural context. It should be noted that, as a hermeneutic device, a clock clearly has a multidimensional set of possibilities, which in turn may fit easily into a number of cultural, multistable structures. In that respect, the clock is a paradigm example of the essential, although non-neutral, ambiguity of technology. And in the examples just noted, while the artifact was transferred, one might almost say the "technology" was not. Or, if the analogue of a hermeneutic device to a text holds, the "text" was transferred, but it was certainly differently read. Only when the entire reading process is also transferred could the clock become the "same" technology.

The Chinese clock example reverberates in an interesting way with the earlier New Guinean example. For the Chinese clock example is one of a successful cultural resistance to a Western "clock culture"—at least, until modern times and until the structure of Chinese society and culture itself changed. But it is also a difference between styles of civilizations. The tribal culture that, in accepting the steel axe, inadvertently accepts a set of dependency relations (such as the New Guinean "cargo cults," which are religions magically trying to entice airplanes to land and give their goods to the practitioners), does indeed get *more* than a steel axe.

Between the extremities of successful resistance to culture-technology and its counterpart quick acceptance there lie the approximate adaptations in which selected ("appropriate") technologies are adapted without total or major disfigurations of indigenous cultures (Southeast Asian examples). I am here reading the situation to be empirically a mixed one and not, as some interpreters would have it, the unimpeded march of Western high-technology culture over and in spite of all cultural resistance.

B. NEOCOLONIALISM AS THE FAILURE OF TRANSFER

The sheer power and concentration of technological and scientific power in the mainly Northern Hemisphere high-technology nations is, of course, an indisputable fact. But it is also an ambiguous fact. It points to a much deeper relation between technology and culture and the embeddedness of the former in the latter. What is not successfully being transferred is precisely the infrastructure necessary for autonomous development—however that development is evaluated.

In a survey of science and technology development by a group headed by Aaron Segal, the contrast between "First" and "Third"

World research and development is looked at to indicate the dispari-
ties between these worlds. Measuring such phenomena is not a simple
task, but gross statistics are at least indicative of the presence or non-
presence of the infrastructure.

The constellations of countries that have such autonomous structures clearly include Canada and the United States in North America,
and the USSR; only to a certain extent the Eastern European countries;
certainly Western Europe, but only fully Japan in Asia (with Korea and
India—and, to a much lesser extent, China—as mid-range powers),
and the isolated exceptions of Israel and South Africa, and Australia
and New Zealand in the South Pacific. All the other some 150 countries at most have partial or selected developments of an infrastructure
or are dependent upon infusions from the aforementioned countries.

The mentioned countries account for 97.1 percent of world research and development dollars and 87.4 percent of the scientists and
engineers associated with such development. This leaves the 150
countries not mentioned with only 2.9 percent of the R&D dollars and
12.6 percent of the cadre of scientists and engineers available for infrastructure development. I read this as a *massive failure* to transfer precisely these aspects of a culture that would support furtherance of
high technology.[5]

I am not here committing myself to an evaluation of whether this
failure is a good or a bad thing. I am observing that what is portrayed
as the successful march of technological culture is not what it is sometimes portrayed to be. Only in the eyes of a precommitted Western
mission can the absence of such an infrastructure be interpreted as a
"lack" (as it is so interpreted by those advocating full cultural-technological transfer).

And even when a country has publicly committed itself to technologization, the results are often partial at best. India is an interesting
example of a Third World country strongly committed to creating its
own science-technology infrastructure: First, India is one of the best
examples of a "successful" colonialism, if by that is meant that many
of the positive values of the colonial power were made indigenous to
the country after independence. Not only has India become a social
democracy—an independent nation-state with a bureaucracy largely
modeled after Britain's—but even its functioning national language remains English. And for at least a hundred years (1850–1950), Western
science and technology were part of this colonial presence. (One
might observe here that the adaptation of some of that science followed patterns noted previously. The extremely impressive astronomical observatories constructed in 1724 in New Delhi and Jaipur, which
contained monumental instruments capable of measuring important

---

5. Aaron Segal, *Learning by Doing: Science and Technology in the Developing World*
(Boulder: Westview Press, 1987), pp. 2–3.

astronomical events down to seconds, were nevertheless primarily motivated by the desire to have the most accurate astrological instruments possible.)

Granted, too, India at its independence found itself following the contradictory paths outlined for it by its two most prominent founders, Gandhi and Nehru. As Brijen Gupta notes:

> The first [goal] has been to meet the challenge of the rising expectations of the Indian people in their most fundamental material and social needs such as food, shelter, health, learning, and work. The other has been to eliminate—or at least, diminish—the dependent ("colonial" or "satellite") industrial and technological relationships with the advanced countries in the North and thus to assert greater economic and political autonomy in the international system. . . . Much to its sorrow, India has discovered that science and technology appropriate for the first end are not the appropriate science and technology to achieve the second objective.[6]

The Gandhian direction, had it prevailed, would have retained more of the premodern culture of India:

> With uncompromising force, Gandhi opposed what he called "mimic anglicism," that is, western clothes, western bourgeois life, western egalitarianism, and western desire to get wealthy and improve one's standard of living indefinitely. . . . So he preached . . . Ramarajya, the revival of the ancient self-contained village community where people would remain poor but not in poverty, because they would not aspire to consuming more that they could produce in the village or obtain in barter from the next village.[7]

On the surface, a drive through the countryside of India shows that village life—with or without the acceptance of being poor—is close to the rule in actuality, yet at the same time there has been a massive governmental commitment to the Nehru ideal of technological modernization.

This ideal, incorporating heavy engineering, scientific research institutes, and electric power still characterizes the main thrust of contemporary Indian policy. The strides have been enormous by multiplication measurements: (1) Since independence in 1948, with an R&D base of 1.1 million rupee crores to the 1,044.9 million of 1984 (most recent measurement), and from an R&D base of 0.23 percent of the 1958 small gross national product to 0.83 percent of a larger GNP in 1983, one sees multipliers not found in any western nation.[8] (2) The number of personnel engaged in R&D activities has grown in the same exponential curve. From the minuscule number of 188 degree recipi-

6. Brijen Gupta, quoted in *Learning by Doing*, p. 189.
7. Ibid., p. 192.
8. Ibid., pp. 199–200.

ents in 1950, India today produces some 2200 degree holders (1985, most recent statistic) from its own sources. Still, these are clearly very small numbers, considering the size of the country (700 million).

Yet the irony is that this is "overproduction," given the infrastructure supporting research towards high technologization. Today, India is effectively an "exporter" of science and technology personnel. Gupta, pointing to these achievements, notes:

> India has not as yet been transformed into a technologically oriented community. . . . There is massive unemployment of technical and scientific personnel, with about 12 percent of recent graduates unable to find productive employment within three years of their graduation. This situation would be worse if the government were not the employer of about 65 percent of all engineering graduates.[9]

India is an "exporter" in another sense as well. When I was in India in January 1988, I learned that every graduate of the Bombay Institute of Technology (one of the most prestigious of India's six such institutes—modelled on an undergraduate version of MIT) in the fields of electrical engineering, computer science, and materials engineering was successful in getting into a United States graduate school. The expectation was that if ten percent of these students eventually returned to India, the success rate would be good. What might be missed in this scenario is that the United States has become an "importing" nation with respect to engineering at the graduate level.

Science magazine has for some time been analyzing the situation in American schools of engineering. A large survey was undertaken in fall 1985, and its conclusions were summarized in "The Impact of Foreign Graduate Students on Engineering Education in the United States":

> The responses provided by engineering chairpersons and faculty indicate that foreign graduate students have assumed an important role in U.S. Engineering Schools, given the shortage of U.S. graduate students. Foreign students were predominantly seen not only as necessary substitutes for U.S. graduate students but as generally satisfactory substitutes.[10]

Currently more than 60 percent of engineering students (in my university, 83 percent) and over 35 percent of assistant professors are foreign, with high numbers of these being Asian, including Indian. (One reason for the dearth of United States graduate engineering students relates to the high salaries available to B.S. degree holders. Our impatient youth apparently want more immediate gratification.) Informally, it is not hard to find our graduate school administrations admitting that

---

9. Ibid., p. 205.
10. *Science*, April 3, 1987, p. 36.

were it not for foreign students, many such graduate schools would fold. We need these "imports" to maintain our own establishment.

In the context here, however, these phenomena are indicators of the failure of technological transfer to Third World countries—even in this case, in a country which has officially adopted a pro-science-technology stance. Of course, in reverse, the policy is a success. The emigrant engineer (and doctor and scientist) keeps our own technological culture thriving. The growing number of co-authors with Indian surnames in science literature is indicative of this reverse success.

What is not being transferred is what might be called a key component of technological culture, the science-technology infrastructure. For high-technology nations, that infrastructure contains science education as part of its culture. Science education may be conceived of as a cultural tool or instrument of high technology. Here, once again, Heidegger was prescient. In his "Question Concerning Technology," he argued for an inversion of the usual model of understanding the relationship between science and technology. For Heidegger, it is modern technology which is "ontologically" the origin of modern science. In a context narrower than this one, Heidegger argued that physics (the foundational paradigm science) is a "tool" of technology as the predominant world-revealing found in the West.

First, modern science becomes technological science in the previously noted sense of being instrumentally embodied:

> It is said that modern technology is something incomparably different from all earlier technologies because it is based upon modern physics as an exact science. Meanwhile, we have come to understand more clearly that the reverse holds true as well: modern physics, as experimental, is dependent upon technical apparatus and upon progress in the building of apparatus.[11]

Conceptually, too, modern physics plays a role as an instrument of technological seeing:

> Modern science's way of representing pursues and entraps nature as a calculable coherence of forces. Modern physics is not experimental physics because it applies apparatus to the questioning of nature. The reverse is true. Because physics, indeed already as pure theory, sets nature up to exhibit itself as a coherence of forces calculable in advance, it orders its experiments precisely for the purpose of asking whether and how nature reports itself when set up this way.[12]

This way of seeing, or this way of having nature revealed, is what, for Heidegger, constitutes the world view of technological culture: "Be-

---

11. Martin Heidegger, "The Question Concerning Technology," *Basic Writings*, ed. David F. Krell (New York: Harper and Row, Publishers, 1977), pp. 295–96.
12. Ibid., p. 303.

cause the essence of modern technology lies in enframing, modern technology must employ exact physical science. Through its so doing, the deceptive illusion arises that modern technology is applied physical science."[13]

What Heidegger has seen is that science (here, physics) plays the role of conceptual instrument within the broader context of a mode of revealing—nature taken as resource well. But the same relationship holds in another way within the cultural matrix. A supportive condition for a high-technology culture is an intensive science education as part of the infrastructure, which supports the enterprise. This crucial factor separates First and Third World developments. Just as the overall infrastructure is lacking, so especially is the intensity of science education, which is important as a cultural motor.

An indicator of this intensity here may be found in the multiple facets of a concern for science education from the official bodies whose interests are fostering science. Within the National Science Foundation budget, for example, is a $139 million item for science education. And while this is a small part of the now nearly 2-billion-dollar NSF overall budget, that single item is larger than the *entire budget* of the National Endowment for the Humanities ($137 million). Out of this NSF program there is much deliberate advertisement for science and science education—including spots on Saturday morning children's television! Similarly, in the media, virtually every newspaper has a science column, television is flooded with science documentaries done in the latest high-technical fashion and, again, most children's channels have regularly featured programs such as "Mr. Wizard." For science journalism, technical writing, and other needs, there are programs geared to keeping both science education and its attainments in the public eye. All of this is part of an intensity above and beyond the educational curriculum, which features heavy science components from kindergarten through college.

While it may be that American science publicity is more intensive than that of other industrial countries, within education itself, science education remains an important part of all technologically oriented countries; there is a virtual competition about comparative attainments. The higher math and science scores of children in both Europe and developed Asia, particularly Japan, stimulate calls for educational reform in the United States. Fear of falling behind in this competition may also be seen in a recent advertising campaign for home computers. Only a few years ago, under a program of the New York Council for the Humanities—on a topic featuring relations of the humanities to technology—my most popularly requested speech was under a title invented by one of the inviters: "Pac Man and Me: Children in the Computer Age." Mental Health Associations across the state called for

---

13. Ibid., p. 304.

family evenings on this topic, with parents turning out in large numbers, in no small part in response to the then-popular television advertising playing upon fears of "Johnny" falling behind if his parents did not rush out and buy a PC.

These vignettes are but indices of the cultural atmosphere in which science and science education are kept a central focus in and for youth. The goal is now often termed "technological literacy" but is primarily an attempt to foster quantitative thinking in all areas. Such is the program in the "New Liberal Arts" sponsored by the Sloan Foundation. This program, targeted for the elite liberal arts colleges (in the belief that if they adopt such a curriculum, the other colleges will follow suit), introduces quantitative thinking and technological relevance into courses as widely different as music composition (computers), history, and the arts.

All of these efforts are part of the designed acculturation process within technological cultures. Without such concentrated efforts, the viewpoints that allow and foster the culture would be lacking. And this educational endeavor stands in just as great a contrast between First and Third World contexts as does the previously mentioned difference in R&D dollars and numbers of research scientists and engineers. Science education is a contemporary equivalent to tribal initiation—at least, in cultural terms. Without that dimension of the wider culture, the transfer of complex technologies that interlink and form systems will remain difficult. These technologies remain what they "are" in relation to the way they become embedded in cultures; without the cultural preparation, the transfer remains frustrated.

All this is recognized in some sense by those middle-range countries that would like to "leapfrog" into more maximal technological development. One often-used instrument for this purpose—perhaps the paradigm "cultural" instrument   is television. Like the radio and the watch, a television is easily introduced, in spite of television being one of the more complex high-technofacts of our time. Its instrumental involvements include electricity systems, a space satellite system (for Third World countries), some distribution and production system— whether government or industry controlled—as well as the complex technology of the set itself. Yet, culturally and perceptually, the television is a "simple" technofact in that what the screen presents is easily and cross-culturally perceived. One of the simplest examples is children's cartoons: with minimal or no dialogue, the characters and plots are easily learned, and even my two-year-old knows the "Winnie the Pooh" videos forward and backward, with anticipations, recited lines, etc. A contrasting technology is the small calculator. Even though this device is small, portable, and, once manufactured, virtually autonomous from any delivery system, it is a hermeneutic device that needs a level of "reading knowledge," that is, mathematics, to be put to "proper" use at all. Again, my two-year old knows very well how to

punch the buttons and make the numbers appear. But he does not
"perceive" the designed significance and—utilizing the essential ambiguity of multiple uses—often holds it to his ear and mimics a portable
telephone. This is to say that Mark already grasps the ambiguity of
technologies in that they can (at least, for the child's pretend world) fit
into a multiplicity of uses. The calculator itself cannot determine either
its uses or the contexts into which it will be made to fit. One could
display it as an interesting perceptual phenomenon—but it is doubtful
that such a fascination would last very long. Even if the television fits
into a different context in developing countries and in Northern Hemisphere societies (where it is present in the home), it is present as a
communal focus and may be found in virtually every Indian village in a
public house.

Television is a cultural instrument in a double sense. It is a concrete instance of a transferred technology, bringing with it a context of
involvements that high technologies have with wider systems, and it is
also a cultural instrument in actual programming and content. This
makes it explicitly a cultural instrument, which must be reckoned with
in equally explicit fashion. The phenomenon of "Dallas" is instructive:
first, this soap opera is virtually international and cross-cultural with respect to all the already industrialized countries. It is just as popular in
South Africa as in middle America and even draws an audience too
large to please the gurus of French culture, thus driving the French in
typical Gallic fashion to create their own more "French" version of the
series.

These, however, are minor variations on technology-culture
transfers, since most high-technology cultures are already highly
"Westernized," (Japan is perhaps the most enigmatic and interesting
counter-example for any too large a claim concerning acculturation at
this level.) Technology-culture transfer in the context of those cultures
that firmly reject many Western values becomes a more serious
problem.

Islamic countries are especially interesting in this respect. On the
one hand, most of these countries are midrange in development and
are striving for technological upgrading. On the other, crucial cultural
elements are in conflict with those of the West. The role and position
of women is particularly sensitive and relates immediately to interna-
tionally available programming (via satellite). Can the role of privatiza-
tion, purdah, and other valued roles for women in Muslim contexts
withstand the cultural transfers via television? This is a highly debated
question in Pakistan, for example. Richard Reeves, in one of his fascinating and perceptive travelogues in the *New Yorker*, found Pakistan to
be a country committed to both modernization and Islamization.
Caught in this double goal, often crucially, were traditional attitudes
towards women. Perhaps extreme were the views of the late General
Zia who, when asked what he did *not* want with modernization, re-

plied, "What we don't want is our women forced out of their privacy. That is what we don't want."[14] When asked about the role of women in a modernized society, Zia went on to say, "We have women engineers. Islam dictates that women cannot be left out of the mainstream. . . . But we are not going to force our women out into the streets. We encourage women to be active within the parameters of Islam."[15] Zia was echoing sentiments of many of the more conservative elements within the society itself. Reeves observed that even medical diagnosis in many rural areas allowed only that men report their wives' symptoms, without allowing any male doctor to see the ailing woman. Another official even went so far as to claim that an underlying reason for the hanging of the previous President Bhutto was related to this traditional value: "Why was Bhutto hanged—really? . . . Because he raised women too high."[16] Of course, later, Pakistan was to be the first country with a woman Prime Minister (Ms. Bhutto). And while she wears traditional garments, gone is the veil of purdah and present is a Radcliffe educated modern. But the current revival of Islamic fundamentalism is the indicator of the countercultural resistance to such transfers.

I am not here implying that "soaps" and other popular television programs are a necessary cultural result of television technology. Although they may be acidic for certain types of cultural puritanism, the soap as such is not at issue. What is at issue is a deeper cultural matter: the pluralism of cross-culturality, which constitutes the ultimate non-neutrality of television culture. I shall return to this theme in another context.

C. "CONTROLLING" TECHNOLOGY

The juncture has now been reached where the popular question about the control of technology can be posed—or, better, reformulated. By placing technology within a cultural context, the two dimensions of the essential, structural ambiguity of technology may be reapproached. At the level of a phenomenology, part of this essential ambiguity was already sighted. The double ambiguity of (a) any technological artifact being placeable in multiple use-contexts, balanced by (b) any technological intention being fulfillable by a range of possible technologies, introduces a certain indeterminacy to all human-technological directions.

Moreover, at the phenomenological level, this structured ambiguity was obtained irrespective of the simplicity or complexity of the technology, at least as human possibility. Medieval castle towers once used as stairways to battlements were later often used for storage bins

---

14. Richard Reeves in *The New Yorker*, October 1, 1984, p. 50.

15. Ibid., p. 68.

16. Ibid., p. 51.

or latrines. Similarly, in the recent debates over nuclear plants, one sees abandonments leading either to conversions or to the establishment of the world's largest "museums" of abandoned technologies. The other side of the ambiguity—in which many different technologies may serve the same purposes—inspires and extrapolates the arms race with unlimited imagined weaponry for the human varieties of destruction. The infinite range of the possible here, unfortunately, fuels the virtually inexhaustible demand for new military-industrial programs.

At the level of a cultural hermeneutics, this problem becomes more complex, but not without the appearance of patterns. Insofar as technologies are cultural instruments and insofar as they are seen to be specifically embedded within cultural matrices, larger shapes of culture-technology also begin to be revealed. This phenomenon reciprocates with the present question over the "control" of technology.

Given this double complexity, now seen at both phenomenological and hermeneutic levels, the first answer to the question of whether technology can be controlled must be a negative one. The reason technology cannot be "controlled" is because the question is wrongly framed. It either assumes that technologies are "merely" instrumental and thus implicitly neutral, or it assumes that technologies are fully determinative and thus uncontrollable. Both extremities are involved in the current debates, but both miss the point of the human-technology and the culture-technology relativities that would reconstitute the debate.

To reframe the question, now in the context of the embeddedness of technologies within cultures, is to see that the question of the control of technology is analogous to the question: Can *cultures* be "controlled"? This reformulation reveals the degree of complexity needed for its answer. Few—except for megalomaniacs historically associated with disastrous results—would quickly answer positively to this reformulation. There is even good reason to see the twentieth-century concern for the "control" of technology as the contemporary equivalent of the nineteenth-century obsession with the "control" of nature. Neither question, in my estimation, is posed properly.

Retrospectively, even in the simplest cases, the question of "control" remains poorly posed. In the tool shop—if for a moment we replace the Heideggerian hammer with a lathe—insofar as the tool-human context is constituted as a relation while the user "controls" the chisel, it is the lathe and its turning of the furniture leg or banister piece that provides the context for the lathe-user's movements. To enter any human-technology relation is already both to "control" and to "be controlled." Once the notion of technology in the ensemble is raised, particularly insofar as technologies are embedded in cultural complexes, the question of "control" becomes even more senseless.

Trajectories of development such as magnification in optics have been pointed to, as have instrumental "intentionalities." Yet at the hermeneutic level, it also has been seen that such trajectories have not always been followed, depending upon the wider and more complex cultural field. The very question of control takes its shape within an implicit, but outdated, metaphysics of determinism. Just as the debate of technological versus social determinism is rejected here, so it should be seen that technologies-in-use do not, as such, determine.

Yet neither are they simply neutral. Technologies, by providing a framework for action, do form intentionalities and inclinations within which use-patterns take dominant shape. This, too, may be seen at the phenomenological level. In an activity highly familiar to philosophers, take note of the following examples of such an "inclination" in the variant uses of three instruments of composition: the dip ink pen, a typewriter, and the word processor.

In this example, consider (a) the editing activity, (b) the speed of composition, and (c) the subtler effect of style. I first became acquainted with the use of the old-fashioned dip pen when my then two youngest children entered an elementary school in Paris. Penmanship was taught as it was for my grandparents, with careful copying, strict instruction, and lined paper. I was fascinated by the uniformity of result; after the children went to bed, I too began to write with the pen. At first, what stood out was the slowness of the writing process (I composed with a typewriter then). But I also discovered that while one's mental processes raced well ahead of the actual writing, (mental) editing could take shape while under way. One could formulate or reformulate a sentence several times before completion. To actually rewrite was painful, and were the object to be a composed letter, it would call for starting over, since there was no simple erasure.

After some experience, accompanied by another phenomenon— the fascination with the actual appearance of the script, whose lettering could be quite beautiful in that the curves and varying scribing could attain aesthetic quality— I rediscovered the "art" of such writing. I could not claim that the use of the dip pen "determined" that I write in the style of *belles lettres*, but the propensity or inclination was certainly there.

Contrasting this experience of writing with that of the typewriter (which was manual), one could see that the ratio of the speed of thought to the actual appearance of the words on the paper was considerably reduced. Editing, still painful, was less so; and a scissors/ paste approach was quite good for drafts. One could not be concerned about the quality of the script, since that was predetermined by the machine. Again, style was not being "determined," although the prettiness of lettering was now eliminated and thus, along with it, one of the features of *belles lettres*. Given the speed of composition, one might argue that a more journalistic or closer-to-speech pattern of style was the more likely "center of gravity" for this mode of composition. Still, it is at best a slight inclination and could be as varied as the self-discipline of the writer.

The electronic word processor poses a different instrumental framework and set of possibilities and is now the favored instrument of many academics. What stands out here is the transformation of the editing process. That has become much easier than with either of the previous technologies, given the ability to reletter and move whole blocks of sentences around. While the speed of composition is but little increased over that of the electric typewriter, it is slightly increased with the electronic as compared to even the electric keyboard of older typewriters.

Precisely because the editing process is made easy, composition now provides a focal temptation. The ease of rewriting becomes a way to see the whole project as more malleable and thus unfixed. Michael Heim has made a similar point in his *Electric Language*. He argues that the computer implies an entirely different notion of the work, of composition, and of language in a revolution not unlike the move from oral to literate cultures. Word processing also encourages the reappearance of what I call the "Germanic tome," the highly footnoted and documented scholarly treatise now made easier by the various footnoting programs favored by scholars already so inclined. (I had actually predicted the "Germanic tome" effect over a decade ago. It has now been confirmed. Bruce Lee, an editor at William Morrow and Co., indicated in *Newsday*, January 29, 1989, that "around 80 percent of the time I am likely to receive a manuscript that is longer than the contracted length of the book. In fact, since word processors arrived, book manuscripts have doubled in length. . . . One reason is the ease in making notes.")

In none of these variants does the technology "determine" the style or the type of composition—but it does "incline" toward some possibilities simply by virtue of which part of the writing experience is enhanced and which made difficult (here returns the magnification/ reduction structure). If one projects such inclinations across many users, the result is closer to predictable at the large-scale social level.

Situating this whole phenomenon, however, is the cultural context. Isolated and specialized modes of writing remain that resist all such technological innovation—for example, the copying of Torahs in the Orthodox traditions, still done on parchment and by quill pen. Ironically, while this religious subculture remains alive, some of its practitioners today operate one of the largest high-technology photo, computer, and electronics businesses in New York City—a fact pointing to technological eclecticism, at the least! Yet the difference between the scroll and the computer is not always a comfortable one, even within eclecticism:

> When the project to computerize the commentary on Jewish law got under way at Bar Ilan University in Israel, the programmers faced a puzzle. Jewish law prohibits the name of God, once written, from being erased or the paper upon which it is written from being destroyed. Could the name

of God be erased from the video screen, the disks, the tape? The rabbis pondered the programmers' question and finally ruled that these media were not considered writing; they could be erased.[17]

If the question of "controlling" technology is misconceived, as I have here argued, this is not to dismiss the crucial issues pointed to in those debates. There are serious issues of which technologies should or should not be developed (there is a serious politics of technological development implied here). The type and degree of technology assessment currently practiced is clearly too minimal and primitive—as well as too controlled by precisely those who need to be "controlled." If the exemplar of risk assessment is itself the highest instance of quantitative thought, there exists here a newer and more subtle form of regulatory conflict of interest. The very agencies whose practices must be assessed insist that the terms of assessment be technocratic in form. In turn, the style of assessment becomes modeled upon the most quantitative of ethical and political theory—some variation of utilitarianism.

There are crucial issues related to the "control" debate revolving around conflicts between the conservation of natural habitats (rain forests, ocean environments, etc.) and development of resources, which forms of technology need subsidy (energy production alternatives), technologies related to labor (intensive versus non-intensive, particularly between developed and undeveloped countries), etc. Yet in a deep sense, while these issues are crucial, they are also primarily political in urgency and shape and are situated within the wider cultural field. Insofar as that is the case, these debates are middle-level ones which must engage a citizenry, hopefully well educated and informed and neither technologically illiterate nor mesmerized by the mystification of "expertise." Yet none of these middle-level issues directly gets at the deeper and broader cultural values that situate the entire field of the debates.

One would like to say that these cultural values are philosophical matters, which they are, but precisely because they are, they are also the least amenable to "control." It may be easy for the philosopher to see that a metaphysics of nature as a resource well is a poor one for enhancing the conservation of even "nature museums," but it is not an easy matter to change widespread cultural sensibilities throughout an entire tradition.

The critique of Western nature domination through science has already been a long theme of philosophical critique, recently joined by feminist criticism of this phenomenon.[18] Precisely because male-female roles are part of a cultural structure, deeply entrenched and long es-

---

17. Michael Heim, *Electric Language: A Philosophical Study of Word Processing* (New Haven: Yale University Press, 1987), p. 192.

18. See especially Sandra Harding's *The Science Question in Feminism* (Ithaca: Cornell University Press, 1986).

tablished, it is difficult to change such patterns. After decades of attack, it remains that, sociologically, the natural sciences, along with engineering, remain dominantly male provinces, resistant to recruitment changes and even more resistant to deep conceptual changes— even with respect to the sorry history of the metaphor of rape or seduction of "mistress Nature" (phrases that occurred in one of last year's Nobel Laureate acceptance speeches!). It seems easier to accommodate males from cultures even more male-dominated than ours into this institutional framework (this is part of the success story with respect to many new emigrants) than it is to entice or accommodate women into that same establishment.

D. TECHNOLOGY-CULTURE EMBEDDEDNESS AS MULTISTABLE

The structured but essential ambiguity of technology has been brought into view, and along with it the phenomenon of variant cultural embeddings. With this positive thrust of the analysis, I have also tried to shift perspectives on some of the debates of the day. This has been the negatively critical aspect of the analysis. To reject the way critical questions have been raised concerning technology, while valuable, leaves hanging the possibility of a positive counter-thesis. Such a thesis is necessarily more speculative and theoretical than what has preceded this juncture, but the basis for such a thesis has ben implicit all along. I may state the thesis in abstract form: Negatively, I have argued that there is no single or unified trajectory to "Technology" (with the capital "T"), that technologies in that sense are not "autonomous," and that the very framing of the question of "control" is put wrongly. Positively, I have argued that technologies are non-neutral and essentially, but structurally, ambiguous. In the relationship with humans and humans-in-culture, technologies transform experience and its variations. Further, I have argued that at the complex level of a cultural hermeneutics, technologies may be variantly embedded; the "same" technology in another cultural context becomes quite a "different" technology.

It is now time to place a name upon that ambiguous structure and describe what I take to be its shape or shapes. The name, following the perceptualist strategy I have chosen for that technology-culture structure, is *multistability*. Its model comes directly from a phenomenology of perception and the polymorphy which perception evidences.[19]

The simplest perceptual example I have frequently used to illus-

19. See my earlier *Experimental Phenomenology* (Albany: SUNY Press, reprinted 1986). There I outline a much fuller theory of visual multistability from which I have here derived a few suggestive examples only.

trate this polymorphy is a phenomenological deconstruction of the famed Necker cube favored by psychologies of perception (Figure 7). Merleau-Ponty also analyzed this ambiguity of the Necker cube in his *Phenomenology of Perception*. My analysis, however, is more deconstructive in that the variations turn out to be much more extensive and variable than those allowed by Merleau-Ponty.

First, a hint at the standard or received view of this visually presented cube: The Necker cube is an ambiguous perceptual object, essentially bi-stable, in which (a) the uppermost part of the figure is seen as the far corner of its top face; but, through a "spontaneous" gestalt switch, (b) the uppermost part is seen as the near corner of its top face, with a second three-dimensional stability. These two variations may switch with each other in the viewer's gaze, in a set of alternations distinct from one another, exclusive but related as three-dimensional appearances of a cube.

A similar example of a gestalt switch of multistabilities has been used by Thomas Kuhn, in *The Structure of Scientific Revolutions*, to describe the phenomenon of paradigm switches in modes of scientific seeing. But Kuhn, as in most psychologies, retains only the example of a bi-stability as sufficient or suggestive for such changes of view.

Phenomenology goes much further in the analysis of perceptual multistability. Its aim is to examine the variations exhaustively to show structural or invariant features. With that search of possibility-structures in mind, such an analysis further deconstructs such multistable objects. I shall only partly undertake this deconstruction here to set the frame for multistable structures—in this case, within visual perception.

To do so, I shall employ a culture-like device to make the variations appear within a perceptual context at once sensory (microperceptual) and yet situated by a "culture" (macroperceptual) in a story. Suppose that, instead of taking the object as the already familiar and sedimented drawing of a cube, I tell you that it is actually a representation of a strange insect in a hole. (See Figure 8.) The six-sided outline surrounding the insect is the outline of the hole. The parallelogram in the center of this drawing is the body of the insect; the remaining lines that project outwards from the parallelogram are the legs of the insect. Usually, after a couple of tries, this new gestalting of the figure occurs easily and can be seen as a two-dimensional variation upon the ambiguous drawing.

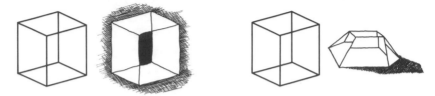

Nor does a phenomenological deconstruction remain satisfied with this newly established tri-stability on what previously was taken as a bi-stable figure. A new micro-macroconfiguration through another story can establish a second three-dimensional variant alternative from and exclusive from the "cube" variation. In this case, the cube is to be taken as a curiously cut "gem" in which the parallelogram is the facet of the transparent stone nearest the viewer (see Figure 9). The six-sided outline is the outline of the gem, but now the shapes around the parallelogram reaching to the external boundaries of the gem are the facets, which three-dimensionally grade off towards the farther-distanced boundary. Here, then, is a second three-dimensional shape, not consonant with the cube shape at all but clearly three-dimensional.

By now, the "secrets" of the deconstruction begin to show themselves. If a reversal of the cube is possible, ought not a reversal of the gem also be possible? That is easily established in a new tale: The gem is now being looked at from beneath; the central parallelogram facet is taken to be "farther" from the viewer than the outlined boundary of the gem, which is taken to be "nearer." Thus a reversal variant on the second three-dimensional shape is attained.

Such a complex perceptual multistability is one of the results of a phenomenology of perception that itself has considerable implication for any psychology of vision, although rarely followed. My suggestion here is that such multistability also may be seen in human-technology relations and even more strongly in the complexities of technology-culture gestalts. To cement this point, I shall return to the navigational examples I have followed and simultaneously develop a historical-anthropological variation upon navigational techniques and extrapolate imaginatively into further technological development.

We have earlier noted how Western navigation took its shape in a mathematical hermeneutic and an instrument-embodied system through which its dominant form employed imagined positions from which to "read" existential positions. Updating that trajectory, one can

see that in the late twentieth century, both the mathematical hermeneutics and the use of instrumentation developed even more highly.
The nuclear submarine carries with it navigational machinery that has
complex displays of digital readouts. These are enhanced by instruments that are, in turn, hermeneutically related to chart positions. In
action, the navigator, with minor exceptions, does not "perceive"
where he is as much as he "reads" where he is by virtue of the instru-
ment-chart learning, which has become sedimented and familiar in a
detailed praxis.

The contrasting set of navigational praxes, which are a different
cultural stability, are those developed by the related schools of ancient
South Pacific navigators. Historically, starting well before the voyages
of Leif Ericson and centuries before Columbus, most of the major Pacific isles had become inhabited, and voyages of interlinkage were undertaken between those distant isles. The difficulty of this navigational
task was considerably more complex than that facing Western navigators in that land masses were much smaller and hence harder to find,
and the distances and spacing within the Pacific were larger; yet these
primitive navigators linked the entire Pacific and peopled even such
isolated islands as Easter and the Hawaiians.

With respect to technology, the early European explorers noted
with awe both the size and sophistication—but especially the
speeds—of the Pacific voyagers. The multihulls, which in high-tech
versions have now broken virtually all sailing records and which have
speeds matching many power craft, were highly developed and capable of long ocean voyages with considerable numbers of passengers and the supplies to sustain them. These craft could easily
outmaneuver and outspeed the slow but heavy cargo-carrying European craft of the same era. (It remains an interesting anomaly in the
history of sail technology that, while the Europeans recognized the
superior technology, they never adapted this to their own purposes.
Here is a technology which did not transfer until the contemporary
times just mentioned.)

What was not technologized, however, was the navigational praxis
of the Polynesian and most other South Pacific navigators. Instruments
were not used, a fact hard to accept by Western interpreters until very
recent times. When I was in Tahiti in 1985, I discovered a treatise on
sailing indicating that as recently as the mid-fifties Western interpreters
were "insisting" that instruments must have been used. The example
given was that of a shell with two holes punctured at the top. The investigator discovered that the sounds the wind made when coming
through these holes varied with speed, and thus he inferred that this
must have been a primitive wind gauge. In fact, it was simply the pen-
dant from a shell necklace; the holes were for the cord.

Instead, the Pacific navigations system could be called a sophisticated and nuanced "Merleau-Pontean" system of micro/macropercep-

tion. First, in it the actual position of the navigator's body becomes the one fixed constant: on board, one did not say one was "going to Puluwat"; one said, "Puluwat approaches." In the entire descriptive framework, the relativity of bodily position vis-à-vis the reference points (rarely abstract; almost always learned islands, wave phenomena, stars, etc.) was consistently maintained. Such perceptual relativity is perfectly consistent and accurately descriptive. From on board, an island does loom ever larger as it appears—one has to presuppose some more ideal or nonembodied position from which to say "I am approaching Puluwat," which is the position so familiar to us that we take it for granted. Similarly, reference points, from the position of the navigator, move. Thus, there was a descriptive language for the ways in which islands moved in relation to each other and for star paths and the other natural phenomena which were perceptually "read" as the system of navigational references.

Islands or land masses were discovered by other perceived significant natural complexes. Long before sighting an actual island, its presence could be detected by (a) stationary clouds sitting over islands in contrast to the simultaneously moving clouds driven by the trade winds; (b) birds returning to roost at night, with different species indicating, through their ranges from island, the approximate distances to land; (c) refractions from swell patterns which "bend" on approaching a land mass; and a plethora of other well-known but difficult to learn phenomena.

One of the most subtle perceptual "readings" had to do with wave patterns. Course direction was taken from the angle of the craft to the dominant swell patterns, and a course could be maintained through day and night by the feel of the craft to swell. Even in local storms, this dominant swell "chord" could be detected in the presence of locally confused seas.

At night, instead of a compass, star paths were noted and hermeneutically arranged. The system of referents was dynamic in that the fixed Pole Star is not visible from subequatorial areas of much of the Pacific. Instead, the parallels of rising and setting stars were followed and embodied in songs which mnemonically recorded the formulae for the different voyages of the repertoire.[20]

Here, then, was a highly complex but noninstrumentally mediated system of navigation, pragmatically and demonstrably successful, relativistic in form, using dynamic constants, and perceptually hermeneutic. It is even anachronistic to call this a "reading" process, since the navigators were preliterate. The system was virtually nonmathematical

20. Observational and instrumental practices of ancient astronomy and navigation are particularly well described in A. T. Aveni, "Tropical Archeoastronomy," *Science*, July 10, 1981, Vol. 213, p. 4505. Archeoastronomy is one of but many hybrid disciplines arising out of science and humanities interests in past praxis.

except for day and night counting, but clearly no number-related geometry was used. The system remained culturally stable, even in vestigial form, to the present (voyages using the system have been reenacted by Hawaiians stimulated by the anthropologists who discovered the systems, most notably David Lewis of Australian National University).[21] It was a counterform to Western navigation.

Again, the purpose here—in spite of the great admiration I have for this system, a system I have failed to learn except in bits and pieces in my own navigational practice—is not to romanticize Polynesian navigation; it is to point up a non-Western but sophisticated cultural attainment that, as long as it was isolated, remained a practical and successful mode of long-range orientation and that contrasts in its key elements with our own mathematically and instrument-mediated system, equally stable and successful. I intend this as an example paralleling that of the cube multistability, in this case as two variants upon the human desire for long-distance orientation across seas stretching beyond immediate horizons. It is an example of differing cultural multistabilities.

Historically, anthropologically, these two systems in more ancient times could be seen as two equivalent systems, one "technological," one not. But existentially and in relation to both perception and the culture-technology matrix, one can imagine a *technological* trajectory which *would* fit the Polynesian praxis. Insofar as the system is "Merleau-Pontean" by virtue of actual bodily position relative to referents, the technological embodiment of such a system would rely heavily upon embodiment relations to enhance, extend, or magnify sensory-bodily perception.

Such a system, for example, would be helped by instruments allowing the navigator to see through fogs or obscure situations. Televisual displays in some aircraft that show runway configurations to the seated pilot are actual contemporary examples of such an instrumentation. Visual displays that do the same thing from "chart position"—often favored in our own systems—are not. For example, in small-boat Loran, a radar signal system popular with sailors, the past fully mathematically hermeneutic devices displaying positions by means of numbers on a video display screen have recently been enhanced by displays showing a chart with a point of light indicating the vessel's position on the chart. This latter development not only has the advantage of the gestalt instantaneity of vision but replaces one step in the reading process: the transfer of digital information to analogue information

---

21. David Lewis, himself an ardent sailor, made South Pacific modes of navigation known through his *We the Navigators* (Honolulu: University of Hawaii Press, 1972) and later in more popular form in the *National Geographic* , as well as in re-enactment of voyages on public television. A very detailed study of Puluwatean navigation was done by Thomas Gladwin in *East is a Big Bird* (Cambridge: Harvard University Press, 1970).

on a chart. But for our Polynesian system, this version of a perceptual hermeneutic process must make its reference to the actual bodily position of the navigator. It would thus have to have a "lateral" display instead of an overhead display.

One could imagine such instruments—and indeed, I would like to have them: for example, a depth finder that would show in three dimensions, perceptually, the configuration of shallow bottoms relative to actual boat position; the previously mentioned fog-penetrating display; or, extrapolating even more imaginatively, an over-the-horizon display of coming islands. In short, the Polynesian system *could* have an instrumental embodiment, an embodiment that relates to its perceptual focus and relativity.

This would be, in one sense, a different technology, or at least, a differently trajected technology, a technology appropriate to the system and structure of the Polynesian cultural stability. But at this juncture, something seems "odd" precisely because, once having discovered multistability, we are already in a position that transforms multistability itself.

The transformation can be suggested by returning to the desired development of imagined navigational equipment. If one system is good, why not two? Or, better, why not have redundant and over-lapped systems? For example, one of the problems facing our own doubled hermeneutic translations between chart displays and actual embodied positions has to do with correlating landmarks. From an embodied position, that spire is to the left of the radio tower. But are these the two indicated points on the chart, or am I somewhere else with a similar marking? A doubled display before one on a console might be even better than one or the other. But apart from imagined instruments which could improve the situation, there is a deeper sense in seeing that *this doubling of systems is precisely where we are!*

To term a phenomenon multistable is already to have recognized it for its ambibiguity and multiple dimensions. Return to the cube/insect/gem drawing: Once the viewer is able to easily and successfully grasp each of the gestalts of the variants—and more, discern the irreversible multiplication of a trajectory in such deconstruction with its relation to related phenomena—that same viewer cannot simply return to the naive viewing previously taken for literal. Here there is an irreversible change of meaning that transforms the Garden of naive perception to something else. If the same applies to culture-technology matrices, there ought to be a similar transformation of this more compelx multistability as well. That too is a phenomenon that lies before our eyes but that needs something of a phenomenological hermeneutic to make apparent.

At the highest altitude, if technological culture is a way of seeing, then it becomes important that we be able to describe what the shape of that seeing is. But at the same time, technological culture, I have been arguing, is not simply one thing. Neither is it uniform nor has its progression across the globe attained either what its detractors fear or what its proponents hope. If, too, a shape of seeing implies nonneutrality, as I have also argued, then it is equally important that the selectivities resulting from this nonneutrality be identified.

To begin this process, I shall return in a more speculative fashion to the phenomenon of technology transfer as related to the types of cultures we have noted.

E. THE VARIETIES OF TECHNOLOGICAL EXPERIENCE

The present march of high-technology culture throughout the globe is a fact that both proponents and critics would agree upon. One result of this march is the decreasing ability of traditional ("primitive") cultures to withstand radical change or possible cultural extinction. These cultures are the most vulnerable to the impact of technologization.

While it is regrettable that this special class of traditional cultures is disappearing, the previous examples have also shown that there is a fairly large mid-range of cultures that have made compromise adaptations, with modern technologies affecting only selected parts (though rarely "chosen" parts, it must be admitted) and leaving large aspects of those cultures intact. Historically, there have been some instances when cross-cultural exchanges involving embedded technologies have been resisted, forgotten, or ignored. The refusal to embed clocks in Chinese culture prior to modern times and the total failure to follow the suggested trajectory of multi-hull shipbuilding by the West are but two prevailing examples.

What I am suggesting is a crude categorization of cultural response to technologies carried by foreign sources to indigenous groups: (1) there are what I shall call "monocultures" of the traditional types, which—virtually no matter what response they take—are overwhelmed by the incoming group; (2) there are middle, compromise adaptations which entail taking selected technofacts into the indigenous culture—these technofacts either may be adapted to a new cultural context or only part of their previous embedded role may be accepted; (3) there are cultures that can successfully resist most of the elements of the incoming group's technologies, although these are rare exceptions to most cross-cultural exchange; and finally, (4) there are cultures that adopt, sometimes even enthusiastically, what is new from the incoming group and modify themselves in some approximation of that group's cultural shape.

The range of previous examples has hinted at each of these categories. If one begins in reverse order, it is clear that Japan, in different historical periods, has exemplified both (3) resistance and (4) the willing adaptation of high-technology culture and, on the surface, even many of its Western accouterments. But it is an adaptation, since both university and industry components are quite different from their Western counterparts (operating by competition and intense elitism in the university and by corporate consensus and group relations in industry). But much of the private sector remains quite traditionally Japanese. Earlier, prior to the opening of Japan to the West, Japan was successful for a fairly long time, in historical terms, in resisting virtually all significant Western influence.

Both these periods, however, are marked by some degree of cultural chauvinism, an ingredient, I would argue, essential to both resistance to and full acceptance of foreign influences. A society must be sure of itself either to reject or to accept what is genuinely "other" than itself. That characteristic certainly marked the Japanese sense of culture during the period of resistance and, in a different way, today.

The same ingredient may be seen in another example less familiar to us. Boorstein points out that one of the feats of Chinese navigation was to have discovered, from the Chinese point of view, most of the coastlands up to the very doorstep of Europe through a series of sea voyages whose primary purpose was simply to proclaim the superiority of Chinese civilization.

These voyages occurred from 1405 to 1433, when they were—like so many reversals in Chinese history—simply cancelled, and China closed herself off from the world for another long period of isolation. But during the life of the ambitious Emperor Yung Lo, expeditions led by the eunuch, Cheng Ho, were made to proclaim to the world the superiority of His Majesty's attainments:

> These expeditions (1405–1433), the vastest until then seen on our planet, enlisted some thirty-seven thousand in their crews, in flotillas of as many as three hundred and seventeen ships. Vessels ranged in size from the largest, the Treasure Ship carrying nine masts, 444 feet long, with a beam of 180 feet, down through the ranks of the Horse Ship, Supply Ship, Billet Ship, to the smallest, the Combat Ship, which carried five masts and measured 180 feet by 68 feet.[22]

Had Columbus been so financed! Or sent to sea in such large ships! Our own history of half a century later might well have been different. His lead ship, the Santa Maria, was smaller than Ho's smallest. Likewise, the Chinese technology of shipbuilding was of superior type for the times:

> Westerners also noted the remarkable construction that prevented water in one part of the hull from flooding the whole ship. Bulkheads, a series of upright partitions dividing the ship's hold into compartments to prevent spread of leakage or fire . . . this design gave the strength and resiliency that made possible the multi-storied ships which dazzled visitors from abroad

---

22. Boorstein, The Discoverers, p. 190.

with their high overhanging stern gallery, from which was suspended a gargantuan rudder with a blade of 450 square feet. . . . Of course, he used the compass and perhaps other directional instruments, along with elaborate navigational charts showing detailed compass bearings.[23]

These expeditions reached all of Southeast Asia, all of the Indian coast, up to the Persian Gulf of recent news. In the Pacific, the expeditions visited Ryukyu, Brunei; and in the Indian Ocean, Borneo and Zanzibar. But what was most unusual lay in the purpose and type of voyage— they were designed simply to proclaim the superiority of the new Ming Dynasty as the world's highest culture:

> The voyages proved that ritualized and nonviolent techniques of persuasion could extract tribute from remote states. The Chinese would not establish their own permanent bases within the tributary states but instead hoped to make "the whole world" into voluntary admirers of the one and only center of civilization.[24]

It is a pity that this generosity—even if motivated by cultural chauvinism—and technique did not last. It was closed off by the succeeding emperor, and later even the building of ships and voyages beyond Chinese waters were made crimes of capital punishment. But in the short time of its life, the Chinese state sent its messages, gifts, and emissaries to the far corners of the southern globe and got, in frequent return, acknowledgment that it was by definition the only truly civilized state.[25]

In the current context, I am suggesting that only what could be called equivalent cultures with respect to power, complexity, and the mentioned sense of self-certainty are likely to be able either to reject or to adopt with alacrity what is met within a cross-cultural exchange of the sort that occurred during the voyages of discovery, whether east to west or west to east. While we are still too ignorant of Asian history, it is at least becoming clear that the multi-thousand-year old cultures of Asia have had a viability different from those of the first category of cultures, those I am calling monocultures.

Examples of monocultures I have cited so far include the inland Aboriginals of Australia, the Inuit, and the Tasaday, to which we could add any number of the disappearing groups of South American tribes in Brazil or Peru. Some similar or generic features mark the situation of a monoculture. First, it is marked by some relatively high degree of cultural isolation or insulation which protects its cultural integrity. The most extreme case cited was the Aboriginal, with tens of thousands of years of isolation.

The relative isolation of the Arctic Inuit was similar; and in tropical

---

23. Ibid., p. 190.
24. Ibid., p. 192.
25. Ibid., p. 193.

contexts, heavy jungle isolation is another form of this necessary separation from many and repeated cross-cultural contacts. Such isolation or insulation provides a kind of natural barrier that creates a habitat or regional environment within which the monoculture can work out its formula of survival and balance. The ways these cultures have been able to come into harmony with their often harsh or arid environments is what strikes the worried late-twentieth-century admirer.

Those attaining this delicate balance, of course, are better known to us than those who failed. To attain a clearer picture, however, the failures or near failures should also be noted. Their bones, including Neanderthal, and artifacts are the stuff of our paleontologies and archaeologies and are the predecessors of the current crop of cultural extinctions, which, to be sure, are more closely linked to societal forces than natural ones.

Monocultures are nevertheless small, strictly population controlled by some combination of cultural and natural controls, and are, in some sense, rarities in history. Not all have had such histories or maintained themselves in sparse conditions. More usual are expansionist or often self-destructive movements resulting in catastrophe. Slash-and-burn agriculture is destructive and can exhaust an environment quickly unless the population remains very small. Desertification, a tendency previously noted in the Mediterranean Basin and today rampant in mid-Africa, is also environmentally destructive. In none of these cases is high technology alone the variable causing the problem (with the notable exception of medical technology, which promotes lower birth and death rates and the prolonging of more older people in a population, thus shortcircuiting the harsh intrusion of the natural environment on a human population).

Indirectly, of course, a stronger case may be made for the negative effects of high-technology culture upon dying or under-stress cultures of the Third World or subcultures within the interstices of industrial culture. The uneven and unjust flow of resources, the elimination of previously arable land, the decimation of wildlife habitats, the exploitation of poor countries for everything from waste dumping to labor use are ills that could be cited as at least associated with the economic infrastructure of technologized culture; but here too these are neither new phenomena nor one do they belong solely to the cultures of the Northern Hemisphere.

What makes the situation more dire and widespread is the larger magnification of power and greater impact of today's technologies in comparison to the past. The introduction of the explosive warhead harpoon in steam-powered modern ships made whaling—even in contrast to the nineteenth-century American industry—a magnitude more powerful and led to the near-extinction of many of the world's largest mammals.

Each of these factors, however, is indirect compared to the acidic effect of high-technology culture upon traditionally isolated, insulated, or unknown human habitats that have been breached. The "other" has appeared, and no culture can be a monoculture in the sense that they can be the only "people." In this sense, technology has allowed contemporary humans to "inherit" the entire earth. The mode of expansion is not merely physical but cultural and is embodied through the various artifacts, particularly those of communication, which link a whole earth into a network. The very conditions of monocultural existence have been breached. Not even some current attempts to create what amounts to "culture museums" will work to save these monocultures. The condition for the stability of a monoculture is analogous to that of a habitat for a specialized species. If that habitat is destroyed, either geographically or culturally, the condition for the monoculture is removed and it either dies or adapts. This is appreciated and understood by what I have called the "culture museum" movement among some anthropologists who would like to preserve such habitats, however unrealistic this might be in the present situation. (The unrealism relates more to the lack of a concentrated and powerful environmental impulse directed towards varieties of conservation than to a negative evaluation. Such a movement would need a deep cultural conversion to become effective.)

However plausible this four-category scheme might be for classifying cultural-technology groupings, it lacks the concreteness both to make a multistable model persuasive and to reframe the questions that I have been criticizing concerning the control of technology. I shall thus revert to a set of examples from recent history to suggest a further line of argument.

I have hinted that contemporary communications technologies are as powerful as they are because of the multiple set of dimensions they incarnate. They can be technologically complex, can or cannot be close-linked to an equally complex set of instrumental involvements, and yet remain hermeneutically simple and virtually cross-culturally available with short hermeneutic learning processes implied.

The examples I shall cite have the advantage also of showing some of the problems with "control" of technology-culture embodiments. Dictatorships—the one form of government that most strongly attempts to control cultures—inevitably seek to control communications media. This is more easily attained in cases of complex, centralized, or close-knit technologies. Large-production newspapers or television broadcasting (until the days of the satellite) are the most susceptible to government control, but no government is totally successful. Here again is an example of the failure of totalization. One means of the defeat of total control occurs through what today are high-technology but decentralized media, which can communicate underground to the dissident populace. A most striking example of such media was the use of the minicassette during the Iranian Revolution. Whereas the

Shah's government was relatively successful in controlling the large media, the propaganda and communication of revolutionary speeches took place nevertheless, through distribution of the small, easily concealed tapes for cassette players. This same indefensibility of a large communications complex was historically preceded by another similar instance in the music industry.

For decades, popular music was controlled by a small number of major recording companies linked, in turn, to a chain of radio stations. Pop music was written virtually by formula and publicized through disk jockeys in the captive chains. When small portable tape recorders became available, however, amateur recordings of folk music, small groups, and other previously non-mainstream music began to be circulated. This was successful enough to transform the entire economic situation of the "majors" so that new companies formed and preferred styles of music became much more diverse than they had been through the previous system. And while a new set of "majors" now has redominated the field, the result clearly is that a larger variety of popular music styles still survives ("country western," "swing," many varieties of "rock"—one of the proliferated styles of the music revolution—"classical," etc.).

While there is a distribution analogy between the Iranian and the Music Revolutions, there is also a disanalogy. The new fundamentalist regime that finally succeeded in overthrowing the Shah in turn sought to control the media, now with *its* new message. While this new regime's attempt at control remains parallel with the recapturing of the music distribution complex by the majors, it is dissimilar in that any resultant expansion of variety is not apparent. Here—at least in principle—another clue hinted at concerning the emerging cultural non-neutrality of contemporary technology may begin to be sighted.

That clue, I shall hold, lies in *the essential pluricultural pattern* that contemporary technology makes possible. It is, however, an inclination, not a determination of technology, as the degrees of restraint against a pluricultural result continue to illustrate. But a *pluricultural pattern is also non-neutral.*

What the totalitarian systems show is that strong, centrally controlled governments can be successful in maintaining a dominant/ recessive system of communications. Insofar as the recessive views cannot be totally suppressed, they can and will exist as mediated through the variable minitechnologies that allow distribution alongside many of the maxitechnologies that remain controlled. Today's array of desktop publishing technologies, cassette recorders, the video camera—all are providing opportunity for decentralized minority expression. Recently, for example, the often-repeated claim that there is police brutality, a claim almost always difficult to prove and usually dismissed by review boards, was given a new twist through the use of a video camera during a curfew dispute in New York City. The results were so dramatic that the police commissioner acted promptly to dismiss and rearrange high-ranking persons in the chain of command. Similarly, television capacities for close decisions in sports events are changing the role and position of referees.

What non-totalitarian systems show even more clearly is the positive possibility of pluricultural results. Television, now with multichannel capacity, has allowed what for a long time has been an "unofficial" second language in the United States—Spanish—to become a major medium of expression through multiple Spanish-language channels. There are less widely broadcast channels in Korean, Japanese, and other languages not even belonging to mainstream Indo-European.

If, however, a pluricultural result has its own shape, its own form of non-neutrality, it is important to show what this inclined trajectory shows. For that reason, I shall once again revert to an imaginative tale.

F. ADAM AND EVE'S CULINARY REVOLUTION

Return for a third time to the imagined non-technological Garden of our imagined primal pair. In this non-technological setting, the culinary scene would necessarily have to be somewhat as follows: First, there can be no technological transformation of food. Thus, un–Lévi Strauss–like, the "raw" will be the only cultural possibility; the "cooked," since it entails fire, tools, and at least a minimal ensemble of technologies, is ruled out.

This leaves us with what may be called a "Garden Core Cuisine." It will include raw fruit and vegetables—at least, those which can be digested and used without transformation; it will thus be considerably narrower even than the Aboriginal diet, which included transformed toxic foods. It could include sushimi-like preparations of raw fish and meat. There would be nuts and the like. Such a diet, while probably healthy, would, for the modern palate (except for the "natural food" fundamentalist), be *boring*.

Lest the example be considered too frivolous, note that cuisines are thoroughly technologically prepared and culturally embedded in practices which themselves have long histories. The earlier revolutions included the use of fire—at first, the prototype of our own barbecue; later, with pots, boiling; and still later, with ovens, roasting. The late James Feebleman once pointed out that ovens and stoves are technologically externalized "stomachs," an observation in consonance with "embodiment."

Furthermore, as previously noted, there developed distinct traditions of both cuisines and the technologies through which those cuisines were prepared. As culinary art became more civilized, there arose whole industries related to porcelain, cookware, and the like entailing complex networks of metallurgical processes, kilnworks, and distribution systems (Mediterranean archaeology is filled with oil and wine urns, still staple food products of the region). Cuisine is, in fact, an excellent example of technological distribution and evolution in a multiple-dimensioned way.

Let us, however, bring the tale up to the present again. Suppose that we bring to our pair some of the multicultural varieties of cuisine that have actually been introduced to our own lands only in the last few decades. First, what was once "Prince Spaghetti Day" has now proliferated into regional Italian cuisines. (As I write this from a Canonica in the hills over Florence, I can verify the soundness of this proliferation. Northern Italian food is differentiated from Southern Italian; Tuscan, from Sicilian.) In Chinese food, once merely the "chop suey" staple of non-Western food for Americans, the choice must now be made between Hunan, Sichuan, Mandarin, etc. To these finer differentiations must be added the whole spectrum of newer and smaller culinary traditions such as Vietnamese, Thai, Afghanistani, Indian in all its varieties, etc. There is even a curious counter-cultural or counter-control turn to recent culinary history. Virtually every new neocolonial conquest or oppressive movement brings in its wake a form of culinary "cultural revenge." From the days of the early civil rights movement, with the popularity of "soul food," to the emergence today of Ethiopian food, each interest of an object group or people sees a cuisine come to the fore.

There is innovation springing from the pluriculinary development. *Nouvelle cuisine* mixes French and Oriental strains, mid-Western food today adopts light cooking techniques, etc. Once our primal pair tastes and becomes acculturated to this variety of cuisines, I would wager that the return to the Garden would become both undesirable and probably irreversible. But culinary pluriculture is an effect of the post-Garden, technologically inherited world.

This tale of culinary non-neutrality is an analogue for what I take to be one of the primary trajectories of contemporary, globe-linked, high-technology culture. This pluriculturality is now an *acquisition of the contemporary lifeworld* and, as such, is a permanent feature of that lifeworld. It is non-neutral, however, in several respects. First, as an acquisition, it serves as a preventer of return to the monoculturality of the Garden. All pluriculturality is acidic for monoculturality. Second, while pluriculturality allows for the persistence of a variety of "cuisines" in the example, cultural traditions in the wider sense imply that none can any longer be taken as simply "true," the only "real people," etc. While there may be a deliberate choice of, or deliberate choosing of, cultivating one rather than another such profile, it is now a self-conscious and thus partially arbitrary choice. And third, this choice will undergo constant testing and revision. If the word processor makes the notion of a "final" text impossible, pluriculture makes a privileged single tradition obsolete.

The shape of the contemporary lifeworld as pluricultural is also acidic to modern and premodern notings of "cores," "foundations," or other ultimate and absolute or even unified integrations. Pluriculture as a technological lifeworld shape *is distinctly postmodern in a limited sense.* (I shall use postmodern in a somewhat technical sense not entirely in keeping with its present use. "Postmodern" in philosophy would be post-Cartesian, post-Kantian, postfoundationalist. In technological culture, it might be emergent postindustrial.) And because it is so, the latent nostalgias of the prophets of technological dystopia are both right and wrong with respect to the future of technological culture.

If I am right about this form of multistability, then the predictions of analytic uniformity (Marcuse), of the victory of technique (Ellul), and even of the sheer world of calculative thought (Heidegger) are wrong. There will be diversity, even enhanced diversity, within the ensemble of technologies and their multiple ambiguities, in the near future.

Yet unless one wrongly senses an inconsistent utopian strand here, it is well to underline the fact that all non-neutrality retains ambiguity essentially neither positive nor negative. If all technologies—and now, here, a hint at an ensemble of technologies—are transformationally selective, one must suppose that the same applies to the only partially imaginative culinary example.

In its perhaps too-imaginative telling, the tale overlooks other aspects of contemporary culinary trends. Could it be that those multiple cuisines—all prepared with loving care by either chefs with long apprenticeships or aspiring young yuppy enterpreneurs—remain under the signs of the last stages of the skilled craftsperson who was doomed by the de-skilling of the Industrial Revolution? That process, too, is clearly evident. From every city and town in the most maximalist technological culture (the USA) the "double arches" now appear only a block from Rome's Spanish Steps and, with the related varieties of fast foods, throughout most of the developed world.

Such fast-food chains are the epitome of the now-computerized assembly-line processes pioneered in the days of Ford and Taylor. There is no skilled chef—the workers are all de-skilled, and one can learn the steps of the process quickly and easily. The marginal persons for employment—adolescents and, more recently, the elderly—become the primary labor sources for the fast-food factory. The market, too, is interesting in that, in Europe particularly, the fast-food operation has become the "hangout" for fashionable youth. Jeans, American food, rock music are points of fascination in a complex that serves as a new "steel axe" or "wristwatch" for youth culture as analogues to the previous examples. In Bacheretto, the Tuscan village from which this is being written, the little village bar to which I must travel each week to make the necessary office call is daily filled with adolescents sitting before the "MTV" channel, listening to songs such as "French Kiss in America."

I admit that the previously mentioned dystopians could take these observations—and particularly given what I suspect is a hidden elitism combined with nostalgia for some folk past—and run rampant with this trend. It cannot be denied that the phenomenon of a youth culture combined with popular culture is a major fact of high-technology civilization. Moreover, it is cross-culturally international and reacts with immediacy from Tokyo to New York to Moscow. I shall return to this strand of the present again.

Youth-Pop culture, however, is but one of the pluricultures to take its place amidst the postmodern spectrum of other traditions. It may be a unique contemporary form of culture, but it is not exclusive and in some respects is limited pricisely by its linkage with an age group.

The culinary example, in both its imaginative and empirical guises, does not fit the model of multistability I have been outlining. Each culinary variant, like the perceptual variations on the cube, has its distinctive appearance and, insofar as genuinely distinctive, is exclusive or alternate to the other variants. Insofar as cuisines are more complicated than the profiles of the visual example, there are possible blends, unique innovations, and the like which may be developed beyond the limits of the cube example. That, however, is an argument for more rather than less "culinary" diversity.

I have also suggested that the postmodern palate is "aware" of this multistability. It has acquired a multiplicity of tastes precisely as a lifeworld acquisition, thus preventing taste from being reduced to any single "core," "foundational," or, in most sophisticated instances, even a narrow range of preferred tastes. Distinctive about this postmodern palate is its shifting ability to appreciate the range of culinary profiles.

One consequence of this ascent into pluripalate status—*it is non-neutral*—is that it will be perceived by anyone still inhabiting a foundationalist stance either as having lost some sense of ground or as dilettante in taste. Contrarily, the person having attained the pluripalate will likely regard the modern detractor as hopelessly provincial and nostalgic for a past long since superseded.

This is the same gain—and the same loss—that every leaving of the Garden entails. It is the price of inheriting the earth. The problem for the contemporary, however, is that in the presence of pluriculturality there is both non-neutrality about this presence and unavoidability in maxitechnological contexts. The culinary tale, however, implies a more serious projection. Multiculturality is a genuine trajectory resulting from the spread of technology across the globe. It accompanies the reach and extent of all contemporary technological invasion into what has been taken as "Western" technology's domination. Multiculturality is an underside of this apparent domination by a history associated with Euro-American technology.

At the end of this second program, then, it appears that a different form of technological-cultural determination has reappeared. Its form is distinctly different from that proposed by the antagonists and proponents of most technological futurism, but its presence is a perplexing one insofar as it constitutes a framework for contemporary choice. Thus a third and final program will be entered, one that concentrates upon the unique curvatures of the contemporary lifeworld.

7. Program Three: Lifeworld Shapes

The task of this third program concerning technology and the lifeworld is distinctly contemporary. Its aim is a partial topography of lifeworld curvatures. Here I have to make a limiting choice. I shall not be so bold or high-altitude as to claim a total topography; yet I wish to catch up, in this analysis, some of the important issues that have been raised concerning contemporary technologies. I have chosen to do a partial topography which reads the lifeworld primarily in terms of an important and relatively new set of technologies, which I shall call image-technologies. These include not only pervasive television, cinema, and photography but also computers—in terms of both word and number capacities—and computer graphics, etc.

This is a middle-level choice for topography, simultaneously concrete enough to remain intuitively tied to embedded technologies and also speculative and broad enough to capture what I believe to be unique or distinctive to the postmodern era. What are the shapes to technological non-neutrality in this postmodern lifeworld?

I have called the present situation "postmodern" not simply to be fashionable but because, in its widespread use, it is a term that an- —nounces an awareness that we are in a transition out of the aura of the modern into what is not yet easily nameable. If this is so, it is difficult either to be sure about the present state of affairs or to make a claim strong about possible, let alone probable, trajectories. Yet there are discernible vectors which do appear; and those, however loosely associated with image-technologies, are of particular interest. They also have the advantage of keeping the inquiry focused upon the praxis and perception, both micro- and macro-, of the lifeworld.

The stance I have taken throughout attempts to avoid both the utopian and dystopian temptations that easily become the sins of many interpreters of technology. Yet in now sketching—in however suggestive a set of lines—the curvatures of high-technology culture, I must recognize that the restraint of mere description will be surpassed. Both one's hopes and one's fears are bound to show through. I recognize that I am a citizen of the very lifeworld being drawn and, as such, I experience both the loves and the hatreds that one always has for one's own country.

Here, too, I must confess the similar love/hate relationship I have for technologies. I share the larger fears of many of the critics of technological civilization that, lacking the necessary conservational ethic and law we need, we may already be close to having irreversibly fouled our own earth—just at the moment we have fully inherited it. That gloomy possibility, while not inevitable, is enough to signal death with a whimper for the human lives on the planet. I fear that possibility more than the other—death by bang through the possibility of global, nuclear holocaust—which is also possible, although not inevitable.

I reject the notion made popular by Heidegger that "only a god can save us." Nor do I have any faith that this could or would happen. In what I shall claim is rather a heightened sense of contingency; we must more than ever see to our own fate, by deeply and even caringly looking after our technologically textured world. While we cannot simply "control" it—since the question is wrongly framed—there are directions that can be taken in crucial interstices that can do some significant nudging.

Many of these, too, belong within the range of image-technologies. But if I am eschewing utopianism, neither do I wish to be dystopian. The contemporary technological world shows distinct advantages, which have so thoroughly become taken for granted that it is easy to overlook those changes, over all previous "worlds." We do not even experience the existential sense of what those changes have been.

The most positive and dramatic change to affect daily life has been in the area of medical and health changes. Gone are a host of diseases: polio, smallpox, and all but vestigially, cholera and diphtheria—the once often-fatal childhood diseases—and others so numerous that the list would fill the page. One could say that this change is also ambiguous—in some crucial areas of the world it has had as a side effect the deleterious effect of population increases that complicate poverty; but in the developed countries, again through the adaptation of birth-control technologies, that problem has also disappeared. One could also say that the religious sense of the contingency of death—a phenomenon daily close to the fourteenth-century European world when the plague wiped out a third or more of the population—has been lost as a religious meaning. But I, for one, do not regret that loss; and while I do think our present attitudes towards death and the institutional ways in which we deal with it clearly need improvement, I would not want to exchange ours for the lost past sense of impending death. In these and many other ways, I admit rejoicing in modernity.

The navigational examples I have used throughout this book also relate to my own positive experience of technology. I am a "religious" sailor in that it is not too far from the truth to claim that sailing is my secular substitute for religion in its ritual, celebratory, and other re-

deeming elevation functions. Here I am unromantically, thoroughly
modern in taste. I am willing to argue that the contemporary high-tech
sailboat is greatly superior to and even more in touch with the natural
elements than any traditional craft. I can appreciate particularly the
smells of tar and wet hemp associated with old wooden boats, and I
like the sounds of their creaking works. But having sailed enough of
their number, I know full well that the capacities of performance in my
own series of fiberglass, fin-keeled, high-aspect sloop-rigged boats is
such that, in both extreme conditions—in which higher-pointing ability
alone makes for greater safety—and in normal conditions—with more
ease of handling and enjoyment and in upkeep, with a much higher
ratio of sailing to repair—it is more than enough to compensate for
whatever nostalgia there may be for the wooden ancestry.

Moreover, the modern high-technology boat, precisely in its capacity to allow oneself to be embodied through it, places one more
closely in tune with wind and water than was so through the insulated
and dampened result in the resistance-to-maneuvering of the older
wooden vessel. A landlubber's analogue here is something like the experience of the road one has through the precision sports car compared to the old American large bomb of a car—good for comfort and
"aiming" but not for enjoying the tortured high mountain roads here
in Tuscany.

Unavoidably, I have now revealed that I can also praise technology in close-to-uncritical fashion, a paean uncharacteristic of the previous restraint. This will not diminish the deeper sense of ambiguity that
must be maintained if critique is to remain genuine. With that, I turn
to the hard and somewhat speculative task of outlining the curvatures
of the contemporary technological lifeworld as seen specifically via the
roles of image technologies.

A. PLURICULTURALITY

The first curvature of the contemporary lifeworld now acquired is what
I shall call *pluriculturality*. It is a lifeform arising out of the use of
image-technologies catching up to cultures. I shall use this neologism
to contrast with the cross-cultural and the multicultural, which are related phenomena not necessarily formed by technological mediations.

There is a persistent *illusion* of neutrality that associates with all
technologies. Its most simplistic form, the neutrality of mere objects in
contrast to motivated uses by subjects, has been avoided here by the
use of the relativistic model that refuses to dissociate subject from object, user from technology. But the illusion can persist precisely because of the multistability we have noted, in which a multiplicity of
users can pick up and use technologies in such different ways. This
occurs also at the level of a cultural hermeneutic, in which technologies get embedded differently. The emergence of pluriculturality will

be seen as the clue to an essential non-neutrality and will serve as a way of deconstructing this illusion at the cultural level.

Pluriculturality is the distinctively postmodern form of the multicultural. It arises in and through image-technologies and is taken into the acquisitions of the lifeworld. Its very appearance mimics aspects of the multistable. Image-technologies are exemplified by a series of technologies including television, cinema, and also computer VDTs with both word and number processing as well as graphics, photography in all its forms, etc. Each of these image-technologies has the capacity to 'reproduce' or *'produce'* "images." (Phenomenologically, I am very uneasy with this language of image, because this usage—now too widely used to simply discard—is steeped in the Platonistic copy and representationalism theory that phenomenology has overthrown. Phenomenologically, an 'image' is itself a "thing itself," that is, a distinctive phenomenon. It is positive, with its own appearance, and does not necessarily belong at all to "representation" but is a distinctive *presentation*. I shall use the language of image, placing it in single quotes to indicate the rejection of its associated epistemology.)

We often take these technologies to be simply neutral "reproducers" of some real thing into an isomorphically produced "image." But no critical user or expert in the medium believes this any more. Indeed, the camera is an excellent example of technological non-neutrality, precisely in the sense that any photo *transfigures* the object which is "taken." The subtler the transfiguration, the greater the illusion of neutrality; but to sharper, late-twentieth-century eyes—at least, in comparison to early viewers not yet educated to transfigurations—this illusion is now seen more as an illusion of representation.

The naive viewer does not miss the importance of being "taken" and transformed by the camera. In the mid-fifties I did a study trip on Southwestern Native American cultures, which included visits to Navajo, Hopi, and Taos settlements. Each group reacted differently to the camera. Navajos were reluctant or refused to have their pictures taken. The explanation given was that the 'image' was really taking something away from the person whose picture was being taken. This had almost "material," although a "spiritual" material, sense. The Hopi, also reluctant, sometimes gave as a reason a belief that such photos would be put to a use which would somehow exploit them—perhaps used profitably for a magazine for which the person being taken would not benefit. The Taos group did not mind their pictures being taken at all—as long as they were paid for the operation! In each case, there was a perception of the non-neutrality of picture taking, although not praised or contexted in a way we might put it.

Although a rigorous phenomenology of transfiguration would be appropriate, it would fit better into the earlier transformations of vision already noted. What I have in mind here is rather that illusion of neutrality arising out of the capacity of image technologies to convey, to picture *any visualizable subject matter whatsoever*—in this special case, the ability to convey but transfigure cultural subject matter.

To convey cultural 'images' was one of the early uses of photography in the now 100-year-old *National Geographic* magazine. It revolutionized the atmosphere of living rooms of many families by bringing pictorially to many provincial folk the variety of world cultures not previously known. It revolutionized what was acceptable in its own tradition beginning in 1896, for example, by depicting bare-breasted women. Placed in the more objective context of authentic photos of authentic peoples, the previously forbidden subject matter—for children, at least—became acceptable. The first visual education about female anatomy for many a young boy at the turn of the century belonged to precisely this convention of the *Geographic*.

What lies deeper than the fascination of the viewer in the now clearly innocent datedness of this practice, however, is the opening to pluriculturality that also begins to show through this use of technology. To bring another culture before one is not a one-way relation. It is an *inter-relation*, even at the seemingly trivial level of magazine photography. The analogue to the culinary revenge of the tale is the 'image,' which begins to transform the previously isolated or insulated home.

Again, there is a relation between the growth of image-technology on a quantitative basis to the qualitative leap occurring within the proliferation. The *Geographic*'s multicultural display was but an early opening to the virtually constant interculturality now seen daily on the evening television news. No informed modern can be unaware of cultural distinction.

Today's informed child, through watching television anthropology documentaries, will know something about phenomena such as potlatches, respect for ancestors, puberty initiatory rituals, and a whole range of cultural phenomena paralleling similar documentary knowledge about black holes and the red shift in astronomy. That child is a better-informed child than the counterpart child at the birth of the *Geographic*. Alongside the esotericism of the anthropology documentary is the often more appealing imagery of MTV—the channel we almost always find still on the set when our baby-sitters leave. Beyond "reproducing" imagery, contemporary television *produces* imagery of its own sort. It creates something of its own subculture. The produced, refracted, and fragmentary 'images' of pop TV are the bricolage remains of photography put into motion but which produce their own variants.

This knowledge through 'images' is not neutral, either. It is as far from the Garden as all the other forms of technologically mediated knowledge we have noted. The contemporary outcry against relativism is yet another form of nostalgic protest. The protesters experience the diminished stature our own history takes in the response to and submergence in the multiplicity of other cultures. But at the same time the appearance of a produced, "image saturated" popular culture, itself filled with multicultural fragments, is also perceived as threatening. Cultural relativism, however, is only a trivialized and often degraded form of pluriculturality. It is also the sign of the *pervasiveness of the pluricultural.*

One reason why the *Geographic*'s excellent early imagery can appear so innocent to us today is that it occurred at the end of the world explorations when it was still possible to sense both missionary zeal and simple Western chauvinism. Each exotic culture 'imaged' could be fascinating but also understood as "primitive" or "savage" and thought to need to be brought into "our" modern era.

Now that our world has impinged upon theirs, there is now a counternostalgia directed at having lost the "primitive" and the "savage." Only their cultural echoes or remains exist. And these are sometimes borrowed, bricolage style, into the songs, clothing, and fashions of popular culture.

If, for a moment, this appearance of pluriculturality is read as the latest variant upon a series of science knowledge revolutions, an interesting set of parallels emerges. Textbook orthodoxy concerning science knowledge explosions usually follows an almost ritual mythos: a new science appears in a theory that in turn changes a major previously held belief, which must be abandoned/decentered—and a new era of informed knowledge emerges.

The Copernican Revolution presumably decentered the geocentric universe, displacing earth and the human to an orbit of the sun, eventually to the status of a minor planet within the Milky Way. The Darwinian Revolution, by relating humans to the primates, presumably decentered anthropocentrism. If the explosion of the cultural is just the twentieth century's latest knowledge explosion, does this mean that the decentering of Eurocentrism is about to occur? Are we at the brink of a decentering of the *Western* cultural universe? That is clearly the fear felt, now being debated in higher education.

Without necessarily endorsing the orthodox views of knowledge explosions, there is an interesting instrumental parallelism in what followed each of the mentioned Revolutions. In each, new instruments were devised; they in turn brought into perceivability what was previously unseen. The transformation of both micro- and macroperceptions followed, which then led to changes of sensibility. Today's array of technologies that make the pluricultural present is the spectrum of image-technologies accompanying this basically only century-old familiarization. Accelerated by the instantaneity of world communications, the present Revolution is now rapidly being disseminated.

Intercultural history is, of course, much older than its image-technology mediated history. A deep reading of our history shows that intercultural exchanges exist at each crucial moment of that history. But with the acceleration of the intercultural during the voyages of discov-

ery—particularly to the New World, then to the Pacific and previously
unknown parts of the East—the fascination also accelerated. The
macroperceptual changes which make for a revolution, however, were
not that easily accommodated or absorbed. Those took longer; a perceptual history reveals something of the difficulty our ancestors had
with perceiving the "other."

Columbus remained convinced at his death that the Arawaks and
Caribs he discovered were but tribes of East Indians somewhere off
the coast of the mainland. The slowness of a cultural, perceptual revolution is also evidenced in the representations of the New World peoples by colonists and missionaries.

Early colonial representations of the New World's indigenous peoples look oddly European—drawn in classical Renaissance style—except for the frillery of feathers and the depiction of clothing (or lack of
it). In North America it was not until the near demise of the original
environments of Native Americans—through the works of Catlin, Rem-
ington, and *early photography*—that realistic representational like-
nesses began to appear. Nor does one have to choose Western
examples to illustrate the same point. Early Japanese representations of
Perry's fleet have the same cultural centric appearance—in this case,
Japanese. In both instances, initial cultural contact is not yet the
pluricultural.

Preceding pluriculturality, in yet another parallel to past knowledge explosions, was the development of a new science: anthropology. Anthropology followed the voyages of discovery by several
centuries. Exploration, with the opening to the West of the peoples of
the New and Pacific Worlds, served as a change in the lifeworld equivalent to that of medieval technology for the birth of science. This
already-established multicultural experience was only later to lead to
the reflective and scientific examination of cultures; but, once entered,
the frontiers of cultural history were to be no less changed in a revolutionary manner than any of the previous decenterings.

No matter how slow the gestalt was to appear, once the explosive
trajectory is launched, it begins a cumulative and quantitative expansion, now multiplied in typical magnificational form by our image-
technologies. The first quantitative result in an "exploded" *canon*. The
previous canon can remain what it is only so long as it is small enough
and select enough to remain sedimented. With at first linear, then horizontal and vertical explosions, the central canon must weaken. This
has already happened and is happening to our own sense of the human, which has become historically and culturally multifaceted:

(a) Beginning only a little over a century ago, the beginning of
human history was first pushed linearly backward. At first, even the
"pre-diluvian" was difficult to perceive or accept, but today it is a
commonplace. Coming first with the Darwinian Revolution, the backward stretching of human time is now over two million years. Indeed,

the latest theory concerning *modern* Homo Sapiens, arrived at through a retroprojection from genetic material, points to an origin in Central Africa ±100,000 years ago. Homo Erectus, clearly humanoid, has an established history of over a million and a half years, older than the oldest reaches conceived of for earth time only a few centuries ago.

Such discoveries in scientific guise usually are portrayed in the standard mythology as the intellectual revolution that finally challenged and demythologized our older and more provincial Western religious notions. But we should not forget that cultures other than ours—most notably the ancient Chinese and Indian—have for millennia held beliefs about the age of earth and humans, albeit hardly in scientific guise, which stretch back in cycles of hundreds of thousands of years.

The decentering of our previous short and linear history is also part of a contemporary reaction. William Bennett's report, *To Reclaim a Legacy* (NEH, 1984), launched a debate about "returning" to a narrow and virtually totally Western curriculum. Such a reaction, however, cannot eliminate the consciousness that now exists concerning this canonically exploded past. Preston Adams, a biologist, made the point well in a response to the reaction in the *Chronicle of Higher Education* (January 23, 1985):

> But, what about *species* literacy? Shouldn't our graduates also have some understanding of the *roots* of human culture, roots that extend backward in the past more than three million years?
>
> Only in recent years have archaeologists and paleoanthropologists begun to understand the world of prehistoric humans in any systematic way. These disciplines have, in the last two decades, brought forth an ever-increasing flood of new knowledge about those humans who:
>
> (a) developed a clay token precursor of writing that worked well for some 5,000 years prior to its rapid transformation into the first writing about 3,000 B.C.;
>
> (b) made the transition, beginning about 10,000 years ago, from living directly off the bounty provided by Nature (through gathering and hunting) to deliberately manipulating the land to grow food (agriculture), thereby setting off the first great cultural revolution;
>
> (c) produced the earliest known art (the magnificent cave paintings and sculptures of the period from about 35,000 to about 12,000 years ago);
>
> (d) buried their dead with reverence, using flowers (some 60,000 years ago, at Shanidar);
>
> (e) solved the problems of living in the shadow of great continental ice sheets (through the use of fire, power weapons—the spear thrower and bow and arrow);
>
> (f) made the very first tools some 2.5 million years ago by sharpening sticks and putting cutting edges on stones (and used these tools to carve out a new ecological niche for themselves and their descendants);
>
> (g) developed language;

and much more, during more than three million years of biological and cultural evolution.

This knowledge, often entailing technological developments in the past as well as having been brought to use through instrumentally embodied new sciences, simply did not exist when what we take to be the core of "Western" civilization took its shape.

(b) There also has been a lateral explosive force regarding the origins of civilization, which now are part of our cultural consciousness. No longer can we rest satisfied with a linear history of civilization going back through Euro-American history, focusing upon the Mediterranean world of the Romans, Greeks, Hebrews, and Egyptians, losing itself in the dim past of the Middle East. Alongside that lineage must be placed the parallel lineages of other civilizations still as extant as ours, including the Oriental ones, particularly of China and India, with which our ancestors interchanged and from whom they borrowed, as other sets of new disciplines have only recently revealed.

Even more, the presumed uniqueness of our lineage must accept alongside itself the developments of civilizational attainments from sources not known prior to the ages of exploration and not fully until this century. The attainments of peoples on our own American continents often equal and parallel those of our Mediterranean ancestors, but because of colonialist prejudice have not been so recognized until this century. The Moche of Peru have taken their place alongside the Inca, Aztec, Tomec, and other South American cultures that had a calendar, writing, astronomy and mathematics, irrigation, and road systems as vast and systematic as did our own ancients.

In the equivalent to our own Middle Ages, Benin art and metalwork excelled that of most of Europe of the same time. Much older than any of these attainments of techné in parallel civilizations is the domestication of grains, which occurred in pre-history in virtually every population area of the globe.

What emerges from modern scientific studies is a mosaic of often interchanging cultures. Indeed, we may not even be too sure about which we should choose as "our" lineage.

To make the point somewhat more cryptically, two science reports concerning archaeological digs made in 1987 struck my mind with a sense of the ironic. Near London, in a dig of a burial plot of about 10,000 years ago, two different settings for human bones were discovered. The one in which bones were laid out in patterns, with artifacts, clearly indicated burial practices. But the other, in the same layer, was of human bones in a random pile, with signs of deliberate bone breakage, especially of skulls, which led the archaeologists to theorize possible cannibalism. A very early find in the American Northwest of about the same era (the dominant theory maintains a human history in North America not to be over 10,000–12,000 years old) confirmed for the first time a new theory about flinty and obsidian fragments. Small chips, once thought to be simply detritus, but then theorized to actually be small microtools (for possible surgical or artistic uses), were found in this site still attached to handles. Thus, while "our" ancestors in England were (possibly) eating each other, the ancestors of the Tlingit and Kwakiutl were sophisticated "high-tech" stone-tool users!

Our linear history can no longer claim even the superiority or firstness of many previously accepted attainments. In mathematics, for example, what was once the province of the Greeks has now become the province of India (where the "zero" was first known to be used) or even the Babylonians (who used the "Pythagorean Theorem" a millennium before Pythagoras). Babylonian mathematics, on a base sixty, included not only most of the calculations of elementary mathematics but also algebra and trigonometry and, linked to a sophisticated set of astronomical observations, was able to make many predictions (not, however, as early as that of the eclipse made by the Chinese, cited previously.)[1]

(c) These astonishing early attainments not only do not belong to our linear histories but are today increasingly being made known to us through new instrumental technologies, which are recovering, uncovering, and finding new histories. This knowledge explosion is yet another result of *technologically* embodied science. In another American example, in 1568 Spanish colonizers moving up the East Coast founded a colony on the coast of Georgia, the Santa Catalina colony, which remained unexamined until the early 1980s. Using an array of sophisticated (and simple) instruments and techniques—such as a power auger (simple), magnetometer, aerial photography, electronic resistivity surveys, and newly developed ground radar—the site has begun to reveal a history quite different from that handed down in the English history that has dominated its interpretation. One cross-cultural outcome, obtained by the new interdisciplinary science of paleoautopsy, "has shown that the Indians suffered a decline in general health after converting from their native ways to Spanish customs."[2]

We are accustomed to such news. It does not shock us, which itself is indexical for the presence of a taken-for-granted pluricultural past and is also an indicator of a now-exploded canon. This knowledge, mediated as in previous revolutions through increasingly varied technologies, opens the interrelation between present and past cultures. In turn, this interrelation cannot help but weaken, possibly decenter, the once-easy culturcentrism of the newer. That is one condition for the rise of pluriculturality in the contemporary.

---

1. For a discussion of similar ancient discoveries see *Humanities*, published by the National Endowment for the Humanities, Vol 7, No. 5, October 1986.

2. Matthew Kiell, "The Magnetomometer That Found Santa Catalina," *Humanities*, p. 22.

This explosion of knowledge has already been felt. Its irony—that it arises from our favored methodologies and through our Western-developed scientific perspective but then places the very history that opened the way as but one "history" alongside many others—is part of what is motivating the current debates about core curricula. The pluricultural awareness arising from the exploded Western canon, in which the ancient pluralities only now become known but take their place alongside the contemporary pluriculturality made available daily through image-technologies, has formed a stratum in our postmodern sensibilities. It is a knowledge explosion not unlike those of Copernicus and Darwin, although no single name can be associated with the discovery of the decentering of "Western" hegemony.

In a symbolic way, perhaps Nietzsche was prophet to this consciousness, and that is one reason for his prominence in so much contemporary philosophical debate. While Nietzsche was clearly aware of the coming death of one era, he was not aware of the configuration of what I am calling postmodern pluriculturality.

To now reaffirm a core curriculum with a heavy domination of Western "texts" must be seen by the postmodernly sensitive to carry a certain cultural arbitrariness. It smacks of a nostalgia for a once-more-comfortable universe that has been enriched both linearly and horizontally. Yet continuing for one more moment to follow the parallelism of the explosions of knowledge storyline, who of us actually experiences nostalgia over the loss of geocentrism? The universe that emerged from the eyeball cosmos turned out to be richer, vaster, and—if the image-technologies depict it successfully—much more beautiful than the medieval paintings that also claim to depict it. I am not quite as confident that most contemporaries would lack nostalgia over the loss of the anthropocentric, prerevolutionary universe as much as they do about the cosmos. That revolution, although over a century old, still retains some of its nostalgic aura, at least in some subcultural areas. But I am rather sure that the *resistance* to what I am calling pluriculturality is healthy and combative, a sign that the consciousness and debate over this issue have really just begun.

There are two ironies in the rise of pluriculturality. The first and broader irony is that the very means by which Western technological culture has spread itself across the globe—its neocolonialism carried by the multiple facets of technological expansion, world economic trade, the reign of science in education, and the intrusion into every previously insulated territory of traditional cultures—has become the very highway upon which the multicultural awareness that precedes pluriculturality travels. But this is part of the sometimes undetected underside of technological non-neutrality.

Image-technologies are communicative, and communications are always implicitly two-way, even in those instances in which there is an imbalance of dominance and recessiveness. Hegel's insights into the master/slave relation and its inversions also apply to communications and image-technologies. That is part of their essential ambiguity, their unique form of non-neutrality. Even if actual two-way or interactive technologies were slower to develop than earlier simply "reproducing" technologies, their current development is but the latest fulfillment of an already extant trajectory.

If the first, broad irony of Western technological dominance is the undercurrent of the pluricultural on the wider cultural scene (evidenced by both the knowledge explosion noted and also the appearance of popular culture as one unique form of the pluricultural), then the second irony is a more focused one. It relates to the specific interpretation of technological science as *a unique and specifically Western development.* If that claim is true—and I believe it is only partially— then the threat of pluriculturality has particular significance to the institutional forms of that technological science. Here we glimpse a different variant upon the Frankenstein myth. The "created" monster who grows powerful and turns on its "maker" is feared within the context of pluriculture in a more specific way.

The felt fear that this is so is often quite explicit in the current debates in higher education. At the very least, the conservatives among the science faculties are quick to join their humanistic brethren with respect to the necessity of re-establishing a core curriculum heavy in *Western* sources and often against too heavy an emphasis upon the non-Western. If Western culture is the motivating culture-shape driving technological science, then to keep it in motion will require continued acculturation to that lifeform. Even the increasing flow of emigrants into this institution must be so educated.

I see much of this debate, current in the reaction against the dilution of Western traditions within higher education, as symptomatic of a fear of a cultural revolution that would weaken or decenter a specifically conceived past. I have cast the scene to this point in terms which take account of the recognition of the threat pluriculture poses to those who would resist it, but its curvature as a positive phenomenon also needs outline.

Unless one wants to be placed in the uncomfortable position of a cultural equivalent to technological Luddism, one must take this latest explosion of knowledge to be potentially at least as beneficial as the past explosions. The array of image-technologies that have transformed cultural perception are bringing the same effects as those technologies that earlier expanded the cosmos and the microworld of the biological. In the explosion of the (Western) canon, there is likely to be as great a decentering as occurred in the previous revolutions. As in the analogue culinary tale, unless one wants to argue that the pluripalate is intrinsically less discriminating or discerning than the palate attuned to a single cuisine, the quantitative result is bound to lead to a qualitative change.

Pluriculturality is, in fact, a *proliferation* of ways of seeing. The worry of those who see it as a threat is a worry over the loss of singular depth, but it is monocularity that gives the least depth in actual human perception. Depth *begins* with binocularity but does not end there. Educationally, we have always affirmed this. To learn a second language not only widens a perspective, it "binocularly" gives us a greater depth of understanding of our own language. This is the case even though, in our past traditions, we have stuck to favoring only those Indo-European languages that were our cousins.

The exploded canon, however, threatens a different kind of proliferation. It may seem *too* diverse for the late modern temperament. That temperament has yet to come to see with what I shall call the *compound eye* of postmodern vision. This compound eye is already instrumentally present in the spectrum of multiple display screens found in settings as diverse as the NASA launch centers and the CBS or NBC newsrooms. Before these compound eyes sits a group of viewers coordinating the separate visions into the "mix" that will be either the evening news or the rocket launch. Ultimately, a pattern and selection occurs, but it is formed out of the multiplicity of individual screens. Risking what I know is bound to be criticized, the compound eye has the advantages of those same eyes in insects; it gives a panorama beyond the boundaries of even the binocular. Through instrumentation, it can transform vision for yet another time in human experience. Such a vision may seem too much for the one-eyed or two-eyed human nostalgically wedded to the face-to-face alone, but the instrumentally transformed world has appeared in this examination of human-technology relations in many other guises. The same now occurs with pluriculturality.

The compound eye as analogue to pluriculturality is already a cultural "image" found in cinema and television. It has transformed vision, but as in all previous analyses, one must continue to be aware of both the magnification and the reduction of all technological selectivities. The compound eye refracts, breaks into vision-bits, produces culture-bits. At first, it is hard to discern what is being seen. There seem to be too many such bits to "read," and the proliferation is very much *unlike* the "reading" of a book.

It is perhaps to be expected that only those who use and learn to use such technologies can comfortably begin to embody them. Also, the early users are admittedly the popular culture users. There is a strand of basically youth culture, international in scope, that has already made such an accommodation. There the indices of the pluricultural are clear in the fluid arts of popular appeal. Fashion, made a respectable academic subject since Roland Barthes, exhibits a rapidity of multicultural borrowings at the level of bricolage. Eastern borrowings in the sixties—Nehru jackets and saris, to which were added African dashikis—were replaced in the late seventies by the paramilitary outfits appropriate to guerrilla movements. In the eighties, higher-tech styles (particularly Japanese) came to the fore.

The arts, always more flexible and quicker to absorb the new than the sciences or the more classical disciplines, are full of pluricultural borrowings at the popular music level. Paul Simon borrows traditional Peruvian Indian tunes for some of his songs; singing styles such as Calypso and Reggae (Caribbean) take their places beside earlier borrowings and adaptations such as the old English tunes echoed in Appalachian folk music. Each of these infusions follows patterns isomorphic with those discerned in technology transfers.

In itself, this popular accommodation to pluriculturality shows neither its full depths nor its possibilities. But those, too, are there. There are far deeper accommodations to pluriculturality that become innovations and even modes of experience, taking on their own life within culture in its narrower, creative sense. In the past, from Africa have come the music traditions which became "soul," the "blues," and above all, "jazz." In the visual arts, the transformations of Benin and Nigerian sculptures in the Musée de l'Homme became "Les trois demoiselles d'Avignon" in Picasso's pluricultural transformation of vision.

The artifactual, too, through the arts and crafts from furniture making to architecture, are indicators of contemporary pluriculture. In furniture, the older tastes for lacquered Oriental furniture never disappeared but have more recently borrowed Japan's futon, adapted again in a transfer-like form to contemporary rooms. Frank Lloyd Wright clearly also drew from Oriental themes for his modern variants, as have other architects from a diversity of traditions. Even the therapeutic skills of the contemporary have not entirely escaped multicultural infusions: acupuncture is partially accepted; meditation techniques—associated with the stress and hypertension phenomena of contemporary lifestyles—continue to have impact. Such multicultural adaptation is so familiar that its presence is likely to be overlooked. It constitutes a present and growing strain of pluriculture within the contemporary lifeworld, now already sedimented as part of all high-technology cultures.

It will be said that such adaptations are not "deep"; and that is often, but not exclusively, true. "Surfaces" are nonetheless positive phenomena. Some of the earliest examples of writing, after all, are no more than lists of goods and inventories; yet without writing as a genuine invention, all the "deep" literature now so called would not be possible. Pluriculture in its present form is an invention of high-technology culture. Its possibilities are not yet fully realized. In the physical sciences today, it could be said that much of frontier science is taking place at interfaces ("surface science," it is called), where much of what is significant occurs at the outer ten percent of materials; their interface processes lead to such things as superconductivity. Perhaps this is a good metaphor for what ought to be similarly explored in the humanistic context.

The pattern that emerges in this shape of the lifeworld is one which (a) sees a proliferation of phenomena (multiple cultural variants), which (b) display multistable, alternative structures, (c) imply a richer field of choice and possibility, and (d) suggest new routes and adaptations. It is the world view of the compound eye and not the narrowed vision of the premodern or even modern vision of the past. It is a "cinematographic" vision, another variation of the contemporary compound eye.

Within intellectual circles the arguments about a transition from modernism to postmodernism are rampant. One leading edge has been the discussion revolving around postmodern architecture. Charles Jencks is a vocal spokesman for such postmodern consciousness. He claims that in the postmodern world, what has emerged is a *cognitariat*, that is, a kind of information-saturated citizen whose very tastes and sensibilities give distinctive shape to the times. He, like those who have reacted against the presence of pluriculturality, recognizes an initial sense of loss:

> Those who have written on Post-Modernism have noted the "loss of authority" which seems so characteristic of our age. . . . No doubt an era of science has cast every ideology into doubt. . . . But the most encompassing trend, it seems to me, is not so much away from belief as *towards an increasing plurality of beliefs*. . . . Mild paranoia is a characteristic of superabundant choice and widespread paranoia . . . [yet] the fundamental shift in mood that the Post-Modern world has brought is a new taste for variety, even incongruity and paradox.[3]

In short, what I have portrayed as a metaphor for postmodernism, the pluripalate, is here extrapolated as a feature of world culture itself. Jencks celebrates it:

> Perhaps the biggest shift in the Post-Modern world is the new attitude of openness. It's not just a taste for the heterogeneity which has brought this about, but also the new assertion of minority rights, of "otherness." . . . Of course many individuals would prefer to limit themselves to a few tastes and develop discrimination within a narrow band, and these loyalties are indeed what makes the overall pattern of pluralism work coherently. But if everybody is limited to a few minority taste cultures, there is still a residual taste for pluralism itself, and for the juxtapositions it entails. . . . The Post-Modern sensibility thrives on dispositions different from its own and recognizes how dull life would be if all took place in the world village. . . . The Post-Modern situation allows its sensibility to be a compound of the previous ones, a palimpsest, just as the information world itself depends on technologies and energies quite different from its own.[4]

3. Charles Jencks, *What Is Postmodernism?* (London: Academy Editions, 1987), pp. 50, 52. I discovered this stimulating essay after I had done the basic work on this chapter and was startled to discover how many of the properties of the *Zeitgeist* we independently had hit upon. Jencks, of course, takes what I am characterizing as a kind of late-twentieth-century "surface" in taste and fashion and makes it much more generally a characteristic of the entire culture.

4. Ibid., pp. 52–54.

Although what has been noted here may at first seem a *defense* of both pluriculture and its "compound eye," that is not my aim. Pluriculture is an acquisition of the contemporary lifeworld. It is one of the ways that world is technologically textured through the proliferation of image-technologies. Moreover, if one of its first results is an international popular culture, that does not yet show sufficiently either the magnificational or the reductive possibilities in full display. It is not at all clear to me whether popular culture is a dominant, emergent lifeform or simply one of the proliferations of culture-fragments that currently describe the scene. It does have to take its place alongside a confusing proliferation of such forms, some as contradictory as the revival of the fundamentalisms, whether Islamic or Protestant. There is a kind of equi-presence of such movements.

What can be said more positively about the popular culture variant among the proliferation of culture-fragments is that it is technologically *experimental*. In an almost naive acceptance of technologies, popular culture and its art forms play with imagery in its *productive*, not "reproductive," aspects. This is so even with as trivial an example as MTV, but more so with the stunning technical effects seen in today's cinema (and I will admit that many contemporary movies are little but technical displays). Cinema has achieved some genuinely artistic and high cultural results in a medium less than a century old. Such uses of image-technology, however, are still too young to be able to project what their trajectories can or will contain.

Pluriculture, both in its acquired form and in its technological embodiment, is closer to a beginning than a fulfillment. Had there been, five thousand years ago at the birth of cuneiform, the educational equivalent of a William Bennett or Alan Bloom, I would surmise that an analogue debate might have occurred then as well. Writing did not seem to be capable at that stage of conveying literature in the full sense but was merely a vast accounting system, a set of bureaucratic records and, as one of its highest attainments for the future of (high) culture, a dictionary. It probably did not seem a particularly promising acquisition of the Sumerian lifeworld. But as a medium, writing, although not in cuneiform shape, was to transform all subsequent culture. Pluriculturality, embodied technologically in the medium of image-technologies, is, however, a *now-acquired vector of the contemporary lifeworld.*

B. DECISIONAL BURDEN

A second and related curvature to the high-technology lifeworld is the heavy weight increasingly placed upon *decision*, particularly conscious decision. Although Heidegger's characterization of the age as one of "calculative reason" is partially correct, it is not yet all of the picture. Even the simple turn to digital instruments enhances such calculative or inferential thinking. In the previous citing of the evolution of clocks, the reduction of time reading to its present "digits" leaves out the durational representation of time (and any "flow" of motion is replaced by the digital "jump" of the Turing machine). This is also the case in digital instruments measuring acceleration (accelerometer) or many other dynamic motional changes. Unlike analogue instruments which display multiple facets of such motions, digital instruments call for subconscious inferences—calculations—to replace more instantaneous perceptions. Such inferences are mini-decisions.

At a higher level, the pervasive institutionalization of ways of reasoning calculatively are part of the "machinery" of many social-political processes. In this examination of a topographical feature of the contemporary lifeworld, I shall detour into the role of life-beginning and life-ending technologies before returning to image-technology. However, these horizonal existential phenomena are far from the only areas indicating the weight placed upon decision.

Part of the institutionalization of calculative reasoning comes in the elevation of the varieties of utilitarianism employed in most forms of risk or evaluative assessment practices. I have already suggested that much, if not most, technology assessment is dominated by such versions of (quantitatively oriented) utilitarian ethical methods.

There is reason for this choice. Most of the practitioners of technical processes are themselves quantitative thinkers. Situations are posed and perceived as "problems" which imply "solutions," the means of which are "rational" (calculative) processes. Those philosophers or ethicists who think in similar terms not only fit nicely into the already extant framework but operate in similar quantitative fashion.

Perhaps the most advanced of the institutionalized instances of such policy-making exists in medical schools and hospitals. Arising out of scarcity issues *created* by new life-support technologies, teams of ethicists were formed to help medical practitioners make decisions about who should benefit in such scarcity situations. The early machines for kidney dialysis were especially part of this early development.

These machines were large, complex, very expensive to operate, and of limited quantity. Yet the diseases treatable through dialysis had more potential patients than the supply of machines could deal with. Since treatment was for critical cases, those who did not receive the treatment were doomed to (earlier—but, in one of the ironies of much high-technology medicine, only *slightly* earlier) death. Doctors, not always reticent about making life-and-death choices, did demur; and medical schools and hospitals soon had committees to make selections and aid choices. This was an early example of "triage," now a common and accepted term used even on entering the emergency rooms of contemporary hospitals.

Ethicists—particularly if they were philosophers—did not easily reach a consensus on how to make such decisions. One popular solution, often argued quite technically in the journals, was basically a lottery, which at least would not discriminate against the poor, the uneducated, or the other likely-to-be-deprived groups. It is ironic that this contrived choice merely transfers the "lottery of life" to a committee!

If the kidney dialysis machine was an early scarcity production machine, since supplanted by massive doses of dollars into the production of multiple miniaturized machines, it was but a harbinger of a well-publicized series of operations and machines that complicate life-support problems even more. Artificial hearts, lung machines, down to CAT scans and NMR techniques for imaging, all raised the technological ante for medical practice and are part of what is today considered normal in the high-tech hospital.

None of this seemingly light-hearted treatment of medical high technology or of institutionalized utilitarians is meant to be simply negative. The crucial decisions dealing with life support need all the help they can get! It is to point up some of the changes that lead to the curvature of high-weighted decisional costs of a technological lifeworld.

Turning for a moment to life-beginning and life-ending situations, I shall display some of these changes wrought by technology upon "deciding life." An opening example in this itinerary was "deciding birth." I pointed out that the contemporary world inverts the ancient world's way of deciding birth. Dominantly, today one chooses when to start life, that is, when to *stop* using birth-control technologies. The border between what would be merely a biological process, allowed spontaneously to take its course, and the technologically embedded cultural process involving explicit decision, has changed. That border has been pushed back "downwards" into the "biological."

When that decision is made and the process does not begin on time or begins with difficulty, the range of options now is also expanded. Fertility treatments, accompanied by more precise diagnoses of where precisely the cause lies or adumbrated by variants once unthinkable, all belong to the contemporary context: *in vitro* fertilization, artificial insemination (both with either the sperm of the spouse or with that of an anonymous donor), transplanted ova (now in a spectrum not long ago unimaginable or only conceptual); choices expand virtually yearly. These techniques within the frame of fertilization technologies clearly are another parallel to the previously noted extrapolation of cultural choice within pluriculture. In a sense, this explosion of choices is a particularized echo of the same former larger-scale phenomenon.

In each of the instances cited, the couple, or even the (so-far female) individual, may *decide* birth or life-beginning. This decision is possible in areas and circumstances changed from all human pasts. If one envisions a line, however vague or sharp, between what could be called the spontaneous and the decisional (between the realm of "nature" and that of "culture"—again I use this distinction with scare-quotes to indicate my uneasiness about such a distinction), then again it is clear that that horizon has been shifted downward into what once would have been thought "spontaneous nature."

The situation is, if anything, even more dramatic in life-ending situations. Not too long ago, death could have been considered simpler, or at least better defined between the spontaneous and the now decisional or controllable. At one time, if one stopped breathing (through drowning, suffocation, smoke inhalation, or a variety of causes), this cessation of a biological process would have been considered a sign of death. Mere cessation of breathing is no longer so considered—indeed, not to attempt resuscitation through techniques as simple as mouth-to-mouth resuscitation or as complex as machine-administered oxygen—would be to be considered negligent.

Similarly, heart stoppage (due to heart attack, shock, hypothermia, and another long spectrum of possibilities), once a "clear" sign of death, is now a matter of decided resuscitation, usually through electric or chemical but even through physical means. Again, the boundary between the signs of life and of death have changed, and today's "brain death" is decidable only through the now familiar intermediary of instrumentation. What is "real" is what is read on the instrument that is more and more the instrumental realism of medical practice. This same phenomenon has also transformed most diagnosis: What the patient says and even the physical palpation of the patient count less than the diagnostic tests, performed by instruments and read by laboratory technicians or other medical hermeneutics. Here the reality seen or perceived through the instrument is taken one step further, although within the realm of the same instrumental realism found in all sciences.

The implication of all these phenomena is that in today's lifeworld I—or my surrogate (wife, relatives, lawyers)—must increasingly *decide* my own death. Such decisions are even planned for in living wills or other increasingly "rational" preparations for deciding death. I have experienced precisely this process in my mother's death only a few years ago:

She had lived to the age of 77, not an extraordinary age today, but under the circumstances, quite remarkable. She had been a rheumatoid arthritic all her adult life and, in the last years of that life, was functionally blind, sclerotized in a sitting position, unable to be mobile or even to fully feed herself, and finally reduced to being cared for in a nursing home in her own community in Kansas.

My father, by this time, was a victim of some degree of Alzheimer's disease and senility; and so, in the Germanic tradition in which I was brought up, as the oldest son, I now had virtually full responsibility for their affairs. As my mother's health deteriorated, circulation problems finally worsened in her legs to the point that, without a double amputation, gangrene would have set in. I was asked to *decide* the operation. I did, affirmatively, but in the full knowledge that the shock to her weakened system would be such that recovery would be risky, even unlikely.

Still, calculatively, death by gangrenous limbs was not suitable. The operation was performed and, for a few days, recovery seemed to be in process; but then she weakened further and soon could exist only by being fed intravenously. A consultation was held with all the relatives, the family doctor, and the operating surgeon present. It would again be my decision as to the course of action.

Several options were outlined: One could help retard the virtual "starvation" process that was occurring if a direct-to-artery operation were undertaken to feed into the neck—but with a failing digestive system complicated by lower-bowel blockage (I was amazed at the progression of technical processes available). After these alternatives were outlined, I realized that the prolongation of some level of minimal biological life could go on for some time but that such measures were, to my mind, too cruel to the patient and, at best, palliative to those who did not want to "decide" a death. I thus "chose" an alternative that had not been mentioned by the surgeon—although it had been discussed with the family physician: proper pain-relieving medication but no further surgical procedures. This was clearly "deciding a death." It occurred two days later.

What has changed in this technological context? Biologically, death remains inevitable. Even the length of life as such has probably not changed very much—life expectancy is longer in the sense that more persons reach horizonal age than in previous historics, but the horizonal age of humans has not itself changed. There remains a border, against which death occurs. But in approaching that border, technological civilization created what I shall call a "Sartrean" situation in which I increasingly must "decide my own death." That is the burden to be placed upon conscious decision.

This feature of the technological age is felt—it lies underneath the rising debates about euthanasia, meaningful suicide, and, at the other end of the spectrum, the alternative means of birth control and abortion. In each of these instances, the very power of decision is felt and seen in its "Sartrean" inevitability. The one choice I do not have is the choice not to make a choice.

Invert the contemporary situation for its ancient norm, and the contemporary situation will stand out in more striking appearance: If, now, in full consciousness and responsibility, I do *not* determine at what stage life-support processes are to be stopped ("no extraordinary means," etc.), then I am "deciding" to place this burden upon others. Do I have the "right" to do this? And, although there is not yet a good reason to raise the issue of "rights" as such, it can be seen that this is a heavy burden to ask of another. Yet some other—not God, not Nature—will have to decide in the widening range of life-supporting boundary situations medical technology has created. This is more than a calculative inclination—it is that; it is also a "Sartrean" existential situation in which I must consciously and responsibly anticipate and "decide" my death in some degree and in some way never before demanded of the normal situation.

In principle, deciding one's death points to the horizonal phenomenon of suicide, suicide in some form as the "norm" for decision. If to refuse to decide is also a decision, its opposite is to decide, in responsible fashion, the boundaries of acceptable contingency. As in the pluricultural field, now in a different aspect in the life-support field, the shape of the high- technology lifeworld proliferates possibilities that, in turn, call for decisions. It raises the issue of heightened *contingency*, but at an end of a spectrum quite different from that of the atmosphere of the plague victim of the fourteenth century. Our being-towards-death takes its vector within the new form of heightened contingency.

Such is a curvature of the contemporary lifeworld. Yet this curvature belongs to the same overall trajectory we have seen in other more trivial or less life-demanding situations. The curvature bespeaks a stretching of both decidability and reversibility. This same phenomenon, evaluated negatively, would be called incompleteness or lack of closure. It belongs in a less dramatic way to the transformation of writing through a word processor. Here we return to image-technology as a patterning process that may be seen in its distinctive curvature.

Michael Heim sees this vector in the spread of word processing, but he gives it a distinctly negative evaluation:

> This superabundance of possibilities [offered through the word processor] is comparable to Nietzsche's description of nihilism as a state of indetermination wherein everything is permitted—and as a result nothing is chosen deeply, authentically, and existentially. Boiler- plate text and reusable fragments haunt all word-processed writing as a constant lure and possibility. Fragments, the experienced user learns, can be used and reused, can be fit in somewhere and without much effort.[5]

Here, in a field quite different from technological medicine, the multi-

---

5. Michael Heim, Electric Language, p. 211.

plication of decidability is perhaps seen even more clearly. Not only is a weight placed upon consciousness in a "Sartrean" way but the lack of closure, the constancy of the dynamic, and the multiplication of options is seen by Heim as threatening. Without necessarily agreeing with his evaluation, one can grant the insight regarding this raising of *contingency* as part of contemporary curvature. Pamela McCorduck, quoted by Heim, echoes the point over: "In other words, electronic text is impermanent, flimsy, malleable, contingent."[6]

These correct but negatively stated insights get their negativity from the implicit model which values only that which lasts, is permanent. Word processing—at least prior to "hard copy"—stands in contrast to that which is complete and closed, a *book*. Thus, word processing is ultimately seen as a threat to the idea of a book:

> Digital writing supplants the framework of the book: it replaces the craftsman's care for resistant materials with automated manipulation; deflects attention from personal expression toward the more general logic of algorithmic procedures; shifts the steadiness of the contemplative formulation of ideas into an overabundance of dynamic possibilities; and turns the private solitude of reflective reading and writing into a public network where the personal symbolic framework needed for original authorship is threatened by linkage with the total textuality of human expressions.[7]

Unlike Heim, I do not worry much about this digitalization as such. First, technologies incline rather than determine. Second, word processing could only have such effects if it were to become the *only* process of writing. Even if it were to become dominant—as it may in some areas—it will doubtless take its position alongside the other multiple variants of texts. Naysayers have been predicting the demise of the book for many decades now, yet more books are published today than in any time in history; and I, for one, do not see such a demise in the works.

Although I have here juxtaposed two very different examples— word processing and medical life-support processes—it should be apparent that the topography is isomorphic. Multiplied decisions, possibility trees, reversibilities within limits—in short, the enhancing of contingency in areas previously more restricted—all belong to the heavier weighting to decision that belongs to the technological lifeworld. I wonder if Heim were to see the same "nihilism" in medical processes, if he would prefer the older and narrower range of decidability.

In the medical example, of course, there remains an area where this malleability does not obtain: It remains that sudden, violent, terminal illnesses, such as cancers of some types, are beyond the bounds of

---

6. Ibid., p. 192.
7. Ibid., p. 191.

much medical malleability, whereas the malleability of word processing is virtually infinite.

Finally, regarding this shape of the lifeworld, decidability with its
weight for consciousness belongs to the same trajectory previously
traced regarding pluriculturality. It is part of the same contingency,
multiplicity, and multistability.

C. MATERIALIZING THE CONCEPTUAL

Modern science differs from its ancient roots by virtue of its technological embodiment. Yet if experiment was born through instrumentation,
it remains something of a historical curiosity that the other root of
modern science—mathematics—remained, until the present, the one
special science which has largely remained *without* a material embodiment. That, too, is now changing.

Contemporarily and almost in spite of itself as institutionalized,
mathematics is becoming a laboratory or experimental science. Historically, of course, there have been mathematical instruments in the history of crude calculating machines. And, less crude, the computer has
become the staple machine which now serves all the sister sciences as
a necessary "mathematical instrument."

It is also regularly used within mathematics but, amongst purists,
sometimes with disdain and suspicion. It remains under the cloud of
suspicion, resistance, and even negativity, simply an application or aid.
But what has become known as experimental mathematics is not,
strictly speaking, a function of the computer's calculational functions.
It is rather the unique way computer hermeneutics transforms mathematical functions into perceptual designs in a return to perceivability
that has created mathematics as an experimental science. Computer
graphics, by transforming number projections through false color into
perceivable patterns, do what calculations do not do.

This new image-technology receives only grudging approval.
Heinz-Otto Peitgen claims: "Experimental mathematics will likely
never be accepted as 'real' mathematics by most mathematicians. But
for many enthusiasts, it has become more than an engaging hobby—it
is, rather, a passion. . . . Such experiments will continue to enhance
our mathematical intuitions in the future."[8]

For those familiar with the history of instrumentation, this reaction
will not appear to be odd—it has been the regular response of all previously "pure" theoretical sciences to technological embodiment.
Church scientists of Galileo's time, perhaps rightfully skeptical of what
the then crude instrument could show, sided with the "purely" theoretical stances of inferential reason.

---

8. Heinz-Otto Peitgen, quoted in "Compositions in Chaos" by Ivars Peterson, *Bostonia Magazine* 59, no. 2 (1985): 35.

Ian Hacking has shown a similar history within biology. The micro-scope did not at first gain acceptance, in spite of the fact that modern biology was born largely through the discovery of the micro-worlds revealed through this instrument. Early microscopes were notoriously difficult to see through. As late as the 1800s, Hacking notes:

> We often regard Xavier Bichat as the founder of histology, the study of living tissues. In 1800 he would not allow a microscope in his lab. In the introduction to his *General Anatomy* he wrote that: "When people observe in conditions of obscurity each sees in his own way and according as he is affected. It is, therefore, observation of the vital properties that must guide us," rather than the blurred images provided by the best of microscopes.[9]

And when we add the problem of many microscopic entities being translucent, we can appreciate what today sounds like obscurantism. Only after aniline dyes could stain the specimen and flint glass was developed to deal with refraction could the microscope attain the near transparency needed to be "seen through."

For the computer to fulfill the same embodiment role as other instruments in the history of technological science, it will have to reveal previously unsuspected and new phenomena in ways parallel to the "artificial revelations" of Galileo's telescope or the microscope. But that is what the hermeneutic graphics do in experimental mathematics. The visualization of topographies of such mathematical phenomena as fractals, chaos, and other random processes has just begun to show such unsuspected phenomena. By turning number patterns into the gestalt instantaneity of perception, the patterning begins to suggest lines of overlap, application, and development not previously suspected. Here, a basically hermeneutic process returns its results to perception. But while its result is graphic, the design pattern comes more from a homologue than an analogue of human patterning.

Early work on fractals revealed "landscapes" which were so like geological ones that these terrains could virtually be taken as representative ones. Philosophically, the graphics began to indicate a "world" both determined and random.

That kind of modeling, of course, returns mathematics to the proximity of its historical use in the sciences, as an application; but it also can become an area of fascination for its own sake, more in keeping with the previously "pure" aspects of mathematics:

> Ironically, the mathematics is proving to be so interesting that many of the mathematicians now working in the field are being led away from the physical applications that originally motivated the studies and away from trying to understand the roots of chaotic behavior in nature. [Devaney,

---

9. Ian Hacking, Representing and Intervening, p. 193.

one of the proponents of experimental math, says,] "We're discovering so many new and interesting phenomena. . . . It's really the computer that generates the mathematical problem. You see something on paper, you try to explain it mathematically, but you can't. So you do more computer graphics, and it goes on like that."[10]

Even though the mathematics used for this first materialization as an instrumental science are simple, the outlook is exciting and not unlike the development of previous instrumentation in science. (Using complex numbers, patterns are generated by projecting the sine and cosine functions. The graphic result provides most of the surprises. The applications and uses will doubtless become more sophisticated.

This materialization of the conceptual, through the instrument of the computer, is a return to perceivability. It is yet another embodiment of a science that, at its modern origins (with Galileo as our symptomatic figure), thought to repudiate the perceptual. Those inventing modern science, echoing theoretically but not practically Greek science, should have recalled Democritus's doubt about the exclusion of the sensory: "Ah, wretched intellect, you get your evidence only as we [the senses] give it to you, and yet you try to overthrow us. That overthrow will be your downfall."[11] In practice, however, modern science incorporated from the beginning, even if only subconsciously, a new mode of sensory perceivability embodied through instrumentation. That instrumentational embodiment today has taken this new form of image technology, which finally makes its appearance in the least "perceivable" of the modern sciences.

This form of imaging, however, is very different from the "reproducibility" technologies of early photography, the presumed lifelikeness of early movie "actualities" or any other of the presumed representations of reality through 'images.' Computer graphics are concocted imagery, a clearly designed *hermeneutic* imagery. They are the analogues of returning writing back towards a kind of pictorial representationalism, a reverse evolution. To transform a number pattern through the artifice of perceptual patterning enhanced by false color, its own stain-process analogue to the early microscope, is a *productive mode of imaging*. Computer graphics stand at an interface between science and art, particularly art in its postmodern cinematographic sense.

Artists have been quick to pick up on this, and computer graphics have already been accepted as media for new art forms. At both of the last Whitney bicentennials, television and cinematographic exhibits have incorporated the abstract art of computer graphics. Last April in Dubrovnik, I saw an exhibit completely dedicated to showing some of the patternings of chaos, fractal, and holographic imaging as productive art.

---

10. Ivars Peterson, *Bostonia Magazine*, 35.
11. Philip Wheelwright, *The Presocratics* (New York: Odyssey Press, 1966), p. 182.

Productive imaging, now established in variant traditions in cinema and television, goes far beyond representation and instrumental realism into innovative fantasy variants now familiar to viewers as well as to those familiar with the literary analogues known longer. Temporality in literature has long undergone imaginative hermeneutic transportation: The flashback, reversed time sequences, refractive times of memory—possibly confused with imagined events—are all the stuff of modern novels. Marcel Proust and James Joyce literally transformed our senses of biographical time before cinema portrayed the same technical tricks in the visually perceivable. Yet the image-technologies of the present return these imaginative productive variations to perceivability. The imagination, too, gets embodied instrumentally. That is part of the shape of the lifeworld and its technological materialization.

There is a social parallelism in both science and art in reaction to such materialization. Mathematicians disdaining experimental mathematics have their humanities counterparts in those who refuse cinema or television art status; yet both forms of instrumentation have transformed and are transforming the previous contexts within which the science or the art has been produced and understood. The genius of the postmodern, however, is not to eliminate any of the previous forms (although they are reduced to equivalent alternations). Each variant becomes *a*, not *the*, choice of expression. To write, one may still pick up a pen—and defend its superiority as an almost sacred form of production in writing poetry, as my neighbor Louis Simpson did recently for *The New York Times*—and one can always turn off the television or refuse to use graphics. Admittedly, one cannot now turn off all televisions or rid the world of computers; any such choices necessarily take their place in an unalterably changed context.

Image-technologies are playing a role in both the sciences and arts in roles that again are analogues to the appearance of pluriculturality. There is a proliferation of choice; there is a return to perceivability; and there is a materialization of both in new forms of embodiment.

D. OSCILLATORY PHENOMENA

A fourth, related curvature to the contemporary lifeworld is what I shall call oscillatory phenomena. These, too, are related to the omnipresence of image-technologies and are the mass responses that exaggerate the actions and reactions of mass movements in travel and communications. A dramatic instance of an oscillatory phenomenon was the emergence of a worldwide student movement in 1968.

This series of events was especially dramatic for me because its strongest occurrence was in France, "les événements de Mai," occurring during my 1967–1968 research year in Paris. I had been attending seminars in Nanterre, where the first student strikes began; had an office on the Boulevard St. Germain where the first student-police conflict occurred; and lived in a quarter with burning barricades.

Whereas the French student strikes were the only ones catalytically to bring a nation into a general strike, simultaneous movements occurred elsewhere in Northern Hemisphere countries. In the United States there was Columbia, the Democratic convention, and Kent State; but also in Berlin, Tokyo, and London students went to the streets and occupied the universities. This was a first, international, and instantaneous oscillation, a mass movement within a mass media context.

What was especially striking, dramatized by the experience of leaving the United States before the cultural and political effects of '68 could be seen and returning after they were in full flower, was the dramatic, oscillatory change that had occurred. Now, twenty-plus years later, in another oscillation characterized by reactionary politics and a revival of fundamentalist religio-politics, one can see that some of the cultural effects of '68 do remain.

When I left for France, only the beginnings of long hair for males, a first-name basis within the university, clothing styles without tie or jacket showed. After return, such appearances were *de rigueur*. While male short hair again dominates, it never again became the universal feature it was only shortly before the cultural revolution of the sixties. Nor have the universities returned entirely to the tweed jacket and tie fashions of before but display the polymorphic spectrum of appearances more in keeping with the postmodern.

That first international oscillation was an indicator of many to follow, respondent to communicational impact (and again, involving image-technologies). That mega-oscillation carried still-extant cultural implications.

One does not have to look to such large-scale events to see the same exaggerative phenomena. Responding to image and information media, one can notice the same oscillations in such items as vocational choices: (a) At one point, only a little over a decade ago, over sixty percent of all entering first-year students at Stony Brook—with not dissimilar results in many other places—declared themselves pre-med majors. Today there is a dearth of medical students, with medical schools worried about maintaining high standards with fewer applicants. We have already noted the even worse dearth of engineering students at the graduate level. (b) Following Watergate, journalism schools were flooded with potential investigative reporters. (c) More recently, it is the flood of masters in business administration. In all these swings there has been an erosion of the more classical liberal arts degrees, at first in the humanities but now with similar long-term erosion in science majors as well.

The student movements were also youth movements, not unrelated to the similar phenomenon of popular culture oscillations. There is a symptomatic location of instantaneity here, but oscillatory phenomena are in no way limited to the volatility of youth.

Technological catastrophe also carries with it an oscillatory response. The triplet of crises previously noted—Bhopal, the "Challenger," and Chernobyl—was followed by immediate and strong public response that led to political results for the technologies involved. The U.S. space program is still recovering from "Challenger" with redesign, investigation, and the restructuring of NASA. The recent decision to mothball Shoreham, a nuclear plant only a few miles east of Stony Brook, was in part due to Chernobyl (and, earlier, Three Mile Island). Business bankruptcies related to Bhopal-like phenomena threaten Union Carbide (the Dalkon shield and the A.H. Robins Co. is another example); all relate to media-enhanced oscillations. These are dampening effects following catastrophic events. They are part of a marketlike but public-image response to communicated news. Oscillatory phenomena are also closely related to the emergence of image-technologies.

These phenomena, now being illustrated as occurring on a large social scale, should not be surprising to anyone now acquainted with the structural selectivities of high technologies. They are a social effect of magnification/reduction selectivities—in this case, those of the mass media. They are further illustrations of the greater powers of contemporary (as compared to any premodern) technologies. They are the media-related cousins of that ultimate in magnificational powers already reached in both weaponry and industrial climatological effects.

Continually before us have been those two technologically empowered forces that have made negative concrete universals actual possibilities. The power to bring about self-induced human extinction through either means has made technological magnification the global power that it is. Of that we are aware, and of that we should have reasonable fears. What, if any, are the dampening effects?

In the area of nuclear warfare capacities, what should strike the beholder as amazing—if that beholder is a technological dystopian—is that for four and a half decades, no major power has reverted to any use of nuclear weapons. This is so in spite of the proliferation of such weapons to probably some twenty nations. Is it the awareness of the extinction oscillation that restrains the megapowers? Certainly nuclear capacity has rendered classical political force partially obsolete and creates a kind of paralysis that has defeated both of the largest powers (Vietnam for the USA; Afghanistan for the USSR).

Instead, in place of a major-power success, there are the small successes and changes occurring within the newer phenomena of terrorism and the politics of these geopolitically sensitive smaller powers. Terrorism is the small-power substitute for the stalemate of megapower, which is powerless. The terrorist act is the propagation of the tape cassette analogue in death technologies. It is uncontrollable in somewhat the same way but continues within the interstices of the larger stalemate. It is a miniaturized technology moving unchecked amidst the megatechnologies that cannot easily or simply be used without awareness of negative consequences. Who said small is beautiful?

More immediately destructive than nuclear capacity has been the megacapacity of industrial pollution upon the global environment. Here the list is long, complex, and only partly isolable for treatment. Atmospherically, lead pollution, acid rain, and ozone depletion head the list. Ocean pollution comes next. North Sea seals, succumbing symptomatically to canine distemper, in all probability are weakened by the high mercury content currently found in their bloodstreams (over 700 ppm). Here, the question is whether public-induced dampening effects can be fast, large, or politically effective enough to counteract the negative impact already known. There are small victories. The most effective to date was Rachel Carson's *The Silent Spring*, which catalyzed action towards the banning of DDT (perhaps a key example of the "feminizing" of science, which I shall point to as a needed corrective). A second, more subtle effect can be noted in signs too small for immediate impact. First, it has been claimed that the membership of the National Audubon Society now exceeds that of the National Rifle Association. If that is true and should a lobby emerge as effective for the first as that the latter now has, one might see a change of direction for wildlife sensibilities. Similarly, and again related to television culture, the proliferation of animal shows for children has clearly changed the field of sensibilities of children in relation to animal life. So successful has this been that some groups are even trying to create children's counter-programming to promote "management" of animals, even to the "humane" use of leg traps. Most of today's children side with the whales (it is unclear whether they also side with the reactor).

The oscillatory phenomenon, in far vaster instances than those cited above, is part of the contemporary curvature of awareness about and coming out of technological society. Its dampening effects are part of the same vector. Who has access to and who controls or effectively uses image-technologies will be a large part of any outcome.

Image-technologies, which have played a large part in each of the shapes of high-technology society, are not the only technologies texturing our world. They are, however, extraordinarily significant as the links of a global system of communications and culturally powerful embodiers of social change. They may be thought of as the nerve system of the interlinked "global village" of the inherited earth.

In each of the four vectors I have traced, I have related their shapes to the emergent image-technologies which are distinctly recent. These technologies, with both "reproductive" and "productive" capacities, are implicated in the full range of human praxes, from science to the arts, and are crucial in contemporary technological texturing. The sub-world created by such technologies is now an acquired dimension of the lifeworld with precisely those characteristics of sedimented familiarity that belongs to all experience using technology.

To some degree there is a notable generational difference in the ability to use and adapt to these technologies and the ease with which initial fascination with them becomes familiar in their social uses. And like all large-scale technologies, the magnificational selectives are such that global effects are frequent and notable. This set of clues recaptures the intuition that contemporary technology is in distinctive ways different from the traditional situation, which remained more insulated and regional prior to networking. At the same time, the structural features noted at the level of a phenomenology and the variants—not expanded—of a cultural hermeneutics remain operative within the human-technology relativities.

8. Epilogue: The Earth Inherited

We are now back to the beginning, the earth as we find it, heavily technologically textured, inherited from the previous generations of humans, all of whom left the Garden. Before turning to concluding recommendations for tending to this inheritance, I shall once again revert to a now contemporary story. In this case the incident is an actual one, deliberately cast, for purposes relevant to the narrative, on one side in a "late Heideggerian" form and on the other with a postmodern commentary. The story is set in the late twentieth century in the foothills of Monte Albano, Tuscany, Italy, now four centuries after the first birth pangs of the modern:

There is a fire of olive and oak, burning in the ancient farmhouse hearth, its smoke curling up the open hood as it would have in the Middle Ages.

(The bricks and stones of the house were gathered from the remains of monasteries and other ruins of the past, not unlike the bricolage dealing in the stones of Rome in the Medieval Period, carted off throughout Europe.)

A man and woman and their young child have just pulled their chairs up to the white marble table upon which is placed a simple Tuscan dinner prepared for the cool October night.

(He, third-generation American, secular Christian, Germano-Scandinavian. She, third-generation secular Jewish, Polish, Russian, Austrian. The child, fourth-generation postmodern bricolage background. Marble, common here, ordinary building material, also occurs in statue of David in Florence.)

First, there is a pasta, freshly prepared with a tomato and mushroom sauce, a sprinkling of parmesan, followed by a homey dish of "patate fritte," all with a

(Pasta, earlier from China, tomatoes, from the New World, mushrooms, truly postmodern international, found in all great cuisines. Potato, again New World, local chianti and mineral water from the local spring in the dale.

the chicken, non-factory produced, is genuinely better than U.S. counterpart. Chianti now harvested by grape combine and fermented in stainless steel vats in local cantina nearby. Water quality controlled for health purposes. . . .)

Dessert is fruit: kiwi and pineapple, not unlike what might have pleased the tenant farmers who once occupied the stone dwelling.

(Obvious imports, Africa and New Zealand. Tenant farmers and padrone system replaced earlier serfs.)

Outside, under the now full moon, the olive trees are ripe and the grapes have already been picked. Fresh figs and almonds remain, along with the incense of the rosemary, oregano, thyme, and sage.

(Olive and vineyards, now several centuries old, replaced forests of previous aristocrats, whose land was deforested to deliberately remove their power by emergent merchant and guild classes. Olives and grapes are among the few products able to grow in now arid climate. Herbs mentioned are typical of same aridity. Tilling is done by track tractor with disk. The world is gathered globally, in terconnected by trade, history, and even the pluricultural cuisine which graces the table. In the living room may be found the television and the stereo; upstairs, the high-tech electronic typer and partial word processor.)

The above scene was, of course, highly enjoyable, centered in family intimacy and an actual event. But the commentary is deliberately cast to de-romanticize the narrative that mimics the settings of Greek temples, peasant cabins and workshops, and the notion of a gathered world in which the ambiguity and complexity of the wider situation is left occluded.

This view adapts the wider angle of the vision of the denuded landscape beyond the Parthenon noted previously, from which the residual romanticism of the here mimicked Heidegger is seen the necessity of adding the unsaid ambiguities relating to the anti-Semitism of many of the peasants and the menacing of pro-fascism. It is important that when gathering is recognized, it be fully and multidimensionally recognized.

The "world" above is not so insulated as its preceding worlds. Were the narrative resumed, one would note that after the fire and dinner, the evening is structured by another set of postmodern choices. After young Mark is put to bed (with stories from his books from Italy, New Zealand, and the United States), his parents have the options of (a) listening to the stereo (Japanese), perhaps to a digitally reconstituted Callas opera (Italian), (b) watching Italian television (which, in the morning, brings CBS news), or (c) retiring to bed to resume their reading: she of Mary McCarthy's *The Stones of Florence*; he, of Iris Orego's *Merchant of Prato*. (Lest the tale be taken as in any way exceptional, note that nearby Florence is filled with dozens of other professors on sabbatical and leave, along with hundreds of students in Italy for their third year abroad, in patterns typical of the high-travel, cross-cultural contemporary world.) The couple chooses to read.

The reading is revealing: The merchant, one Francesco Datini, who bequeathed to posterity 503 files worth of his letters, papers, and ledgers, brings the late medieval period to us through detail and person. In the late 1300s, at the time a pope inhabited Avignon and was constantly in conflict with the Northern Italians, at the very eve of the Renaissance, one could see the glimmers of the coming modern era. This entrepreneur, all too willing to trade in anything, selling armor and religious articles to all parties (even warring mercenaries on all sides), traveling most of the civilized world of importance at the time or having business outposts as a harbinger of the supernational conglomerates of our own and in an age already noted as having established its own technological revolution, reveals to us his fears in his lifeworld.

There is the plague, from which he barely escaped and to which he lost most of his family twice. There are the religious wars, the last of which finally disenfranchised him from Avignon, allowing him to return just in time to see his guardian foster-mother before her death. There are the robbers, and dangers of travel. And there is famine, which occurs with surprising regularity, even in the Italy of the fourteenth century.

Less than a hundred years later is to be born, illegitimately, as was so frequent at the time, one Leonardo; born in 1452 in Vinci, a village just around the shoulder of Monte Albano, only four decades before the New World was to be discovered. He was to become the very symbol of Renaissance polymathism. He was clearly a herald of the technologically embodied science that was to emerge from the Renaissance. Opportunistic in the extreme—not unlike his Tuscan predecessor, Datini—da Vinci offered himself to a series of wealthy aristocrats and warring lords. He wrote to Ludovico il Moro an offer to build innovative battle machines:

1. I know how to build very light, strong bridges, made to be easily transported so as to follow and at times escape from the enemy. . . .

2. I know techniques useful in invading a territory, like how to drain water out of moats and how to make an infinite number of bridges and covered walkways useful . . . for such expeditions.

3. Item, if in the course of an offensive, the height of an embankment or the strength of a site should preclude shelling, I know techniques for destroying any fortress or other stronghold not built on solid rock. . . .

4. Whenever the shelling fails, I will invent catapults, mangonels, traps, and other unusual and marvelous instruments. . . . [1]

This engineering science is as wedded to the "military-industrial complex" as any Eisenhower ever dreamed of! (We have already noted that Galileo, yet another century later, followed the same path.)

In the twentieth century the same thing happened. Very shortly after the discovery of fission, Werner Heisenberg, seeking to recoup his reputation, wrote secret letters to the War Ministry of the Third Reich and later organized a conference, the proceedings of which were titled *Probleme der Kernphysik*, in which he proposed nuclear-powered submarines, battleships, and a super-explosive which was to launch the Nazi attempt to build an atomic bomb. The difference with da Vinci lay mainly in the now corporate and multiauthored structure of twentieth century science and technology.[2]

The birth of Renaissance science is a birth within technological garb and institutionally wed to the same sources of finance as today's Big Science. It is only the nineteenth-century successful myth that has convinced us that it was ever otherwise. Here, too, is the doubled relation to technology that occurs at the birth of *modern* science. It is embodied in instrumental technologies, but embedded in a matrix of engineering and linked to the largest-scale patronage available.

In a last look outside at the lights of the Florentine Valley, the contemporary man of the tale reflects upon his worries: There remains part of the haze over the valley, which can be seen in some degree every day. The plague is gone, replaced by a much slower process in atmospheric pollution. The Germans to the north have begun to realize that their two intense loves—for forests and for automobiles—have now reached contradictory straits. They have initiated actions concerning pollution controls and unleaded gasolines, even against the delicacies of Common Market politics. Farther north still, the Swedes have decided that one cannot always have both the whale and the reactor and have chosen to phase out their reactors (although, until recently,

---

1. Marco Cianchi, *Leonardo da Vinci's Machines* (Becocci Editore, 1988), pp. 17–18.
2. David Cassidy, a historian, is now doing a definitive history of Heisenberg during the War years. He provided me with copies of previously classified materials authored by Heisenberg, including correspondence and the *Probleme der Kernphysik* (Schriften der Deutschen Akademie der Luftfahrforschung, 1943).

their westerly neighbor, Norway, was one of the countries still killing whales).

Famine is unheard of in these parts now, and Italy has surpassed England in production and gross national product. But the man knows that in the South, in the former colony of Italy—Ethiopia—there is still famine, abetted by the very indigenous government that replaced the colonizers, by desertifying farming practices, and by the lack of sufficient aid due to world political tensions.

The moral of the tale is clear enough: Although nostalgias and romanticisms may—in small doses—soften our harshest views, they can also obscure and sometimes *dangerously* obscure issues. What is more strongly needed than either of these medications is a deeper sense of the ambiguity of technological civilization in both its negative and positive vectors, that is, its heightened sense of contingency.

This sense of heightened contingency is itself a legacy of our current immersion in technological texture. It is part of our inheritance of the earth, a dimension of the non-neutral way in which we have received and taken up that inheritance. So the harder question is how we will care for and handle that inheritance. It could be squandered; it could be conserved; and it might even be increased.

What should we do to opt for the latter two alternatives? In part, the project of this book is a preliminary recommendation in answer to precisely that question. It is preliminary in a philosophical sense as a serious attempt to come to grips with the more adequate grasp of the shape and structure of a technological civilization. It is in this sense a very classical philosophical approach in outlining a framework of understanding. I have tried to be critical and yet encourage neither an outright rejection nor a blind acceptance of our inherited state of affairs. But in that very process, it is philosophically inevitable that there will be implicit evaluative perspectives that simultaneously emerge from within the inquiry itself. It is to these that I turn for the closure of this project.

Stewardship Recommendations for the Inherited Earth

A. TO CONSERVE THE EARTH

A major sub-theme of this project has been the recognition that the high magnificational powers of technology now carry geological impact force. Combined with what remains a dominantly expansionist ethic towards nature, my first recommendation must be a worldwide conservational ethic. This is, to my mind, the most sweeping and urgent need in response to the currently often negative relation that high-technological civilization has placed upon the environment.

While the dominant Western values cannot escape blame, these are not the sole source of the problem; the blame needs to be distributed much more widely than that. J. Donald Hughes, in his scathing critique of the lack of an environmental ethic in *all* the parent civilizations surrounding the Mediterranean in ancient times, sees part of the blame in the abandonment of an even more ancient animism. Whatever the modes of escape from animism were—Greek, Hebrew, Roman—the results were the same for the Basin; it is not clear that the more ancient forms of religion and culture were, overall, any better.

Other non-Western cultures have all too frequently been destructive of the environment in practice. One of the native American tribes of the Iroquois Nation, out of the belief that animals had speech and would gossip about any secret or war party travels, would systematically kill any animal *en route* lest the animal tell of their whereabouts. I have previously noted the systematic extinction of wingless birds throughout most of the Pacific by the westward march of the various tribes of my favored navigators. Even among traditional societies, it all too frequently appears that delicately balanced societies such as the inland Australian Aboriginal or the Inuit are exceptions rather than the rule in rigorous conservational practice.

What has allowed this to occur—to the present day—has been the richness of the earth and smaller populations to burden it. Slash-and-burn societies could continue to exist, but only in a jungle that could recover after their travels and only on condition that the population of those peoples with such destructive practices remained small enough for that recovery to occur.

On the other hand, our own historical record has not been invariably negative. Heidegger was correct about one aspect of European peasant culture. It did care for the land as soil. René Dubos has pointed out in *Wooing the Earth* that one result of improved European farming practices after the invention of the plow and discovery of rotation farming was an actual building up of a nutrient-rich soil. In France, that improved soil became the property of the tenant farmer who did the upgrading and could be taken with him should he move. Similarly, one of the few areas of the world to seriously reforest, after virtual deforestation, has been New England in the United States. There is more forest now in the northern New England states than at any time since the colonists first arrived.

Two centuries ago, 70 percent of the land in Rhode Island had been cleared of the deciduous forest that once covered it almost completely. The primeval forest had been transformed into agricultural land by the original white settlers. During the late nineteenth century, however, the less-productive farms were abandoned and trees returned so rapidly that less than 30 percent of the state remains cleared today.[3]

One side effect has been the successful migration back or reintroduction of many previous species of animals not seen for over a century or more. In Vermont, my own vacation home now is a site for wild turkeys, fishers, and even the occasional coyote that has adapted to forest life.

While the overall news remains bad—deforestation is clearly proceeding more rapidly than reforestation—conservation movements of some success are occurring in the midst of the world's most populated sectors. Europe, for example, is the most populated (persons per square mile) geographic area on the face of the earth, and yet when actual attempts are made to control or reverse negative effects, there can be small successes:

> In the London area, the Thames River has long been extremely polluted, as attested by Michael Faraday's much publicized letter to the *Times* in 1855. The abundance and variety of fish had started to decrease almost two centuries before . . . and only eels survived in certain areas by 1855. As of 1976, however, there were eighty-three species of fish in the estuary, and even salmon were caught in London for the first time in approximately 150 years.[4]

To distribute blame, however, is not the same thing as letting one's own traditions get off with mitigating circumstances. I here join Martin Heidegger, Lynn White, Jr., J. Donald Hughes, René Dubos, and many others who have laid a good deal of the responsibility at the door of some of our dominant cultural and religious beliefs. These must be modified if the conditions for a genuine conservation ethic are to be found. Those beliefs include (1) the belief that earth is primarily a resource well for privileged human use; and (2), closely linked, the belief that there is a large and valuable significant gap between humans and the rest of the animal kingdom such that our precedence is justified in virtually all cases.

Contributing to these beliefs is a long and interwoven history that includes not only one strand of Judeo-Christian earth dominance beliefs but also the strand of the interpretation of earth arising out of earlier Greek atomism that reduces earth in a different way to a kind of random material (which could, in combination with the former belief, become the way to subdue that resource-well earth). If the dominant

---

3. René Dubos, *The Wooing of Earth*, p. 33.
4. Ibid., p. 41.

view is both reductive and domination-prone, I remain unconvinced that any revival of ancient animism or its relatives is an option or is desirable.

Langdon Winner, in his usual insightful appreciation for ambiguity, has posed the dilemma well in both the title and experience of *The Whale and the Reactor.* Can we have both? In an experience which Winner likens to the epiphany of Henry Adams in "The Virgin and the Dynamo," but obversely, he finds the choice crystallized:

> I looked out over a vista that sent me reeling. Below us, nestled on the shores of a tiny cove, was the gigantic nuclear reactor, still under construction, a huge brown rectangular block and two white domes. In tandem the domes looked slightly obscene, like breasts protruding from some oversized goddess who had been carefully buried in the sand by the scurrying bulldozers. . . . At precisely that moment another sight caught my eye. On a line with the reactor and the Diablo Rock but much farther out to sea, a California grey whale suddenly swam to the surface, shot a tall stream of vapor from its blow hole into the air, and then disappeared beneath the waves. An overpowering silence descended over me.[5]

There is an asymmetry in the question. We *can* or could make substitutions for the reactor. There can be other technologies; and there can be other, multistable contextings of technologies; but we *cannot* replace the whales. This applies equally to the nearly half the species under threat in rain deforestation practices. In an article, "No Dinosaurs This Time," a case is made that

> "Earth's biota now appears to be entering an era of extinctions that may rival or surpass in scale that which occurred at the end of the Cretaceous, some 65 million years ago." This gloomy and more than a little startling opinion was expressed by Paul Ehrlich, of Stanford University, at a recent meeting on the dynamics of extinction. Rampant development, including urban, agricultural, and forest-felling, is the cause of the impending collapse, he says.[6]

And while I like the "jungle" versus "desert" arguments of the biologists who are today arguing that if we lose this richness of both animal and floral variety, we will also lose a lot of potential "solutions" to the world's future pharmacy, it remains only a pitiful variant upon still seeing the earth as resource well. Surely the earth is richer than a potential pharmacopia, but can we see this without returning to the dark animism of our ancestors?

---

5. Langdon Winner, *The Whale and the Reactor: A Search for Limits in an Age of High Technology* (Chicago: University of Chicago Press, 1986), pp. 165–66.

6. Roger Lewin, "No Dinosaurs This Time," *Science,* vol. 221, September 16, 1983, p. 1168.

To see a need for a world-wide conservation ethic—with legal support for its particulars—is far from unique as a recommendation. It has been repeatedly called for and recognized as a need. Nor is it sufficient to complain that, so long as the present structure of nation-states, multinational corporations, and needs of developing nations pertains, no significant action will result. Deeper, but harder to deal with, are the cultural/philosophical issues which set the context within which the debate occurs. As long as nature is seen as reductively instrumental to a humankind elevated above all its biological neighbors, in the mode of our current dominant mythology with all that follows from this mythology, there will probably not be sufficient motivation for change. Neither instrumental arguments nor aesthetic ones—particularly since in modern culture aesthetics always takes second place to utilitarian positions—will sway sensibilities.

The catch is that any larger gestalt switch in sensibilities will have to occur from *within* technological cultures. If the cultural hermeneutics of this book are suggestively correct, then no larger, external culture of sufficient strength exists to persuade us of its superiority. There is no Chinese fleet on our horizon. This is merely the secular rejection of any saving "god" as likely or possible.

The larger scene, then, appears to be gloomy regarding any immediate global persuasion concerning the enactment of a conservation ethic. But within the interstices there are small, positive indicators which might be enhanced. First, networking the entire globe through various extant technologies is a condition for any genuine global change. That networking remains a factor simultaneous with technological "dominance." Second, as the phenomenon the pluriculturality demonstrates, one unpredictable effect of any linkage is the two-way flow of cultural influences along that network. One effect already detectable is the weakening of certain traditions *of* precisely the Western values which have prevailed. Pluriculturality does not indicate anything like the rise to dominance of some specific non-Western cultural model. Although many would like to see the adoption of various Eastern counter-beliefs as a balance to the present Western dominant strain, nowhere do I see that as a likelihood. This weakening, while a condition of any more ultimate change in sensibilities, also does not necessarily portend either a betterment or necessarily a worsening of the situation. The emergence of the strong postmodern strands of popular culture are, at most, ambiguous and as yet too difficult to measure with respect to a more particular cultural formation.

Within the wider, largely youth cultures of postmodernism there are a few smaller indications of positive directions. I have already cited the animal empathies, to which should be added a much stronger awareness of ecological interconnectedness appreciated by most television-educated children of today. Another positive factor that—to me—seems to be missing or to have largely disappeared from consciousness through television education is the past form of negative stereotypes associated with certain animal species. (The "big bad wolf" now more clearly belongs to fairy tales, less to wolves and their habitats. This is also a positive factor in genuine science education as disseminated through mass media.)

These empathetic and ecological forms of awareness have as a secondary effect the weakening of beliefs in a justifying gap between humans and the animal neighbors. In addition, help from precisely the materialist side of our past has also emerged. The very sciences that have de-animized (and de-anthropomorphized) both the animal and the human now are gradually reducing the sub-experiential aspects of the human/animal gap. This is occurring at the levels of both behaviors and biology.

We are now able to recognize such a phenomenon as "animal culture," learned and transmitted behaviors akin to human historical cultures, at least among higher animals. Wolves—once despised—are recognized as having particular societal patterns transmitted to their progeny through social behavior. Whales change their songs (do they have a postmodern popular culture?) between years of their migrations, indicating something much stronger than innate repetitions. Tool use, primate patterns much closer to our behaviors, even proto-language skills and, below the level of behavior, genetic overlap (of 99 percent between humans and primates), all lessen the biological distance between us and our animal neighbors.

In spite of this modification of the gap between us and the animals, there is yet to be strong evidence that we are willing to modify our own behavioral stances towards animals. This morning's news, however, carried the National Academy of Science study recommendation against *any change of law* regarding control of the laboratory use of animals in experimentation related to research. I am not here arguing that animal rightists either have or do not have a point against the science establishment; I am pointing out that at the same time that the sense of our neighborly boundaries are changing, the practices of established science are being defended as necessary. (The NAS very often acts in a fashion similar to the AMA in medicine—this should come as no surprise.) It will be seen, however, that so long as the animal kingdom remains part of "resource well" nature, the diminishing of distance can make little difference by itself.

If there are small signs of changes in the human/animal set of relations, does the same apply to our concepts of nature? The dominant, reductive sense of nature remains pervasive. Can our concept of nature be deinstrumentalized without at the same time returning to some form of animism? Are there any internal indicators that would modify the present dominant attitude toward nature? The answer is not an easy one, nor are there any easy escapes from the present climate. Obviously, any raised awareness of the human interlinking with the

ecological system of the earth is positive. And, apart from the attacks
upon the "Gaia Hypothesis" which still associate with anthropomorphism, the concept that sees an interlinkage between biological and
non-biological dimensions of earth seems both likely and plausible for
the future of the understanding of the earth. At the very least, to see
technological civilization as a kind of biologically activated "geological" force should by now be clear. With or without the metaphor of
"Nature telling us something," the self-reverberation of negative environmental actions are more widely known.

Still, there is no postmodern, hopefully richer model that can
compete for wider recognition here without recalling some nostalgic
association with older religious or pre-scientific understandings of
nature. Again, I am not arguing against those who would introduce
elements of such past culture fragments—that is one of the very tendencies of postmodernism that I have argued characterized pluriculturality—but bricolage remains theoretically dissatisfying.

Thus, as urgent and as clear as the establishment of a global conservationist ethic enacted in an agreement of international or multinational laws may seem, it also seems unlikely that such an
enlightened result will occur very immediately. Here, then, strategy
must remain interstitial and piecemeal.

B. DEMYTHOLOGIZING (AND DEMASCULINIZING) TECHNOLOGICAL SCIENCE

In its critical sense, hermeneutics has a *demythologizing* function. In
the context of the programs of reframing our understanding of technology, I have repeatedly taken a modified view of science as the conceptual tool of technological culture in its high-technology sense. And,
if to change or reshape a tool has effect, then one might hope that a
significant interstitial action could be directed toward this end. The
history and philosophy of science has indeed undergone radical
change in the last two decades, and a host of critics has begun to
reach a consensus upon several important aspects of understanding
and interpreting contemporary science. I would term the critical dimension of that consensus one that aims at demythologizing exaggerated myths concerning science.

The rise of science education has been seen to be essential to the
maintenance of technological culture; science is the factually most
dominant set of disciplines within academia; and there is a quasi-
religious belief that adheres to science as socially salvific; these
functions are all deeply entrenched in high-technology culture. Demythologization has as part of its task balancing—and in some cases,
reducing—the distortions that can occur through current myths about
science.

In its largest outline, the emergent consensus agrees that science must be seen as *one* of many human intellectual and cultural activities; and as one, it must take its place (more democratically) among its peers. That is not the role it currently plays in any of the advanced technological countries, particularly our own. Especially within education and even more specifically within the large research universities, science—and its associated "applied" relatives—plays the overwhelmingly dominant role. That is evidenced both economically and politically. At the national level, one *small* indicator is the relative size and support structure between science and the arts and humanities, as seen in the National Endowments. The now nearly $2 billion NSF budget—scheduled to be doubled by the present regime—dwarfs the approximately $135 million of the NEH and NEA respectively. Add the support of industry and the sometimes hidden, but often more overt, support of the military, which supports the large part of many academic research budgets (see graph), and one sees that there is no peerage here at all.

HOW MILITARY DOLLARS ON CAMPUS DISTORT SCIENTIFIC RESEARCH

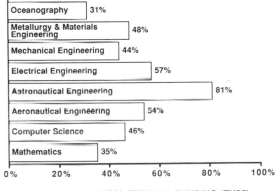

DOD SHARE OF TOTAL FEDERAL FUNDING (FY85)

SOURCE: Jonathan B. Tucker, "Scientists and Star Wars," in Union of Concerned Scientists, *Empty Promise* (Boston: Beacon, 1986); and William Hartung and Rosy Nimroody, "Star Wars: Pentagon Invades Academia," *CEP Newsletter* (Council on Economic Priorities, January 1986), Table I.

Then, when one adds the nearly $8 billion budget of the National Institutes of Health, which supports much of the biology establishment, one can see that between disciplines there is a much larger disparity of support than between the relative sizes of David and Goliath. Nor has one ever heard of Goliath simply willingly giving up the higher ground.

Most Davids who take on Goliaths end sadly, but in stories they sometimes win a point or two; and stories are what historians, philosophers, and literary critics are best at telling. It is in that arena of interpretation and critique that reframing can occur. Those critiques take several forms. I shall not here rehearse all of them but focus only upon several of the more pertinent and newer forms the critique of science as institution is taking.

My first example of the emergent demythologizing consensus comes from a new twist in a *close-up* controversy between science and philosophy of science, which appeared in a battle over a possibly commissioned commentary, "Where Science Has Gone Wrong," in *Nature* magazine, one of the most prestigious of the British science journals. It is strong evidence that Davidian storytelling does make itself felt.

By now in the nineties, it is apparent that a wide-fronted intellectual revolution spanning many disciplines began to take shape in the early sixties. In the humanities, this was the period of the upsurge of a whole spectrum of "theory," the birth of many methods now called the "New Scholarship." It included the impact of postwar phenomenology and hermeneutics, structuralism, revivals of neo-Marxian and neo-Freudian methods, and later, feminist and deconstructive techniques as well. But at the same time, although almost in insulated form from the above largely Continental trends, there emerged what I shall call the New philosophy of science. In North America, this new interpretation of science focused in Thomas Kuhn's *The Structure of Scientific Revolutions* (1962). In Britain, a less radical reformulation of science interpretation was developed by Karl Popper.

What can be called the Old philosophy of science tended to interpret science in a noncontextual way, focusing almost solely upon its logical, propositional, and rational procedures. As a result, science as an institution was made to appear to be ahistorical and acultural—and, most relevant for an examination of technology, clearly distinct from and often divorced from its technologies.

Each of these features was to be challenged by variants upon the New (history and) philosophy of science. These new movements called for an interpretation of science which made it a *situated*, a *contexted* phenomenon. Without here going into the particulars of this "paradigm shift" concerning science interpretation, I merely note that we have had two and a half decades of contestation within history and philosophy of science circles in which the proponents of the Old philosophy of science hurled epithets of "relativism" and "ideology" at the proponents of the New. But in spite of that, it is fair to conclude that the New philosophy of science has pretty much secured its place on a wide front—indeed, many in the sciences now interpret their own work in Kuhnian and post-Kuhnian fashion. The New philosophy of science, Davidian in its initial hermeneutic, had—if not slain—at least wounded the older giant.

What I am calling here the '*Nature* controversy' is a part of a reaction to that battle—but in a new guise. The authors do not resurrect the Old philosophers of science as their defenders, even though they call for something close to an early Positivist stance, but instead, attack the presumed results of the battle with respect to what they perceive as eroded public support for science endeavors. This motivating issue was the decline of public spending upon science in Britain, particularly in the universities, throughout the Thatcher era. The authors noted that the public climate for spending on science began to cool in the seventies. Part of the blame, even the deepest part of the blame, for this decline of support, the authors laid at the door of the New philosophy of science: "our objective is to identify and endeavor to combat what we consider to be the most fundamental, and yet the least recognized, *cause* of the present predicament of science, not only in Britain but throughout the world."[7] Inveighing against the public broadcast of programs discussing issues of objectivity in science and entailing critiques by New philosophy of science positions, these Bennett-Bloom twins noted:

> These were attacks against *objectivity, truth* and *science*. . . . At least in Britain, the repercussions of these mistaken arguments are already happening. Scientists in other countries are duly forewarned.
> We shall refer to these erroneous and harmful ideas as the epistemological antitheses—the (un)philosophical positions which are contrary to the *traditional and successful* theses of natural philosophy.[8] (italics mine)

The tone is obvious enough, and the polemics of the article match the worry. Note, of course, that the worry is directed to the diminishment of what could be called damage to the mythological aura of science in the realm of public belief; but equally, a high-priestly tone comes through as well:

> It is an objective of this article to refute these ideas, and argue that the *correct epistemology* is indispensable in any serious and responsible scientific work. *For what is really at stake is nothing less than the future progress of our civilization.*[9] (my italics)

The fervor of these defenders of 'correct epistemology' was naturally strongly challenged, not only by philosophers but by scientists who saw the merits of both the rigorous logic practiced by many philosophers and the insights provided through the situating of science in historical, social, theoretical-contextual, and paradigmatic fields. The controversy raged for over a year; and in the end, the authors eventually replied to their critics thusly:

---

7. T. Theocharis and M. Psimopoulos, "Where Science Has Gone Wrong," *Nature* 329 (October 15, 1987): 595.
8. Ibid., p. 595.
9. Ibid., p. 597.

A frequent and passionate objection was to the dogmatic way in which
we express some views. We never cease to be amazed by the frequency
and vehemence of this objection—how can anyone seriously say with
such passion anything at all if one does not have complete confidence in
what one says? The reason why so many people are horrified by
dogmatism is obviously because many instances in the past caused much
harm. But in all these cases harm was done because what people were
dogmatic about was not true. If what one is dogmatic about is true, this
dogmatism cannot possibly be harmful.[10]

Had such a response occurred within literary circles, it might well have
been taken to be a subtle satirical parody of an ancient voice. But my
reason for including this almost ludicrous example—no matter how
prestigious the magazine in which it appeared—is that it reveals something about a mythos which is often taken to surround the science-
technology establishment. From a postmodern consciousness, such
dogmatic convictions about presumed universal and objective truth
justifying a holy war against errant opinion could hardly be perceived
as other than archaic. Or else it is the emergence of yet another form
of twentieth-century fundamentalism.

There is also something of a postmodern irony here—for the article is directed at the loss of budget support for British, particularly British academic, science. Yet the USA, led by an equally reactionary
government, did not seem to be affected by the dissolution "caused"
by the New philosophy of science. Rather the build-up of R&D dollars—especially linked to military and industrial contexts—was dramatically increased. In an assessment of Reagan era support for science
and technology, the AAAS reports a nearly doubled support level (see
Table 1).

Should one conclude from this that the overall 85 percent in-
crease in U.S. science R&D occurred because our departmental sec-
retaries and their staffs failed to read the New philosophers of
science, whereas the British cabinet and its staff did—and followed a
presumably implied conclusion that science funding should be cut?
The entire correspondence in *Nature* avoided the obvious concerning public disillusionment related to negative effects of much large-
science policy. Radioactive pollution of the Irish Sea, acid rain, urban
smog, etc., are all well-publicized issues in Britain as well as here.
And while purists can argue that this is merely the wrong policy for
the *application* of science, the public often sees science-technology
as a single, large unit.[11]

---

10. Theocharis and Psimopoulos, *Nature* 333 (June 2, 1988): 309.

11. Steward Richards, *Philosophy and Sociology of Science* (New York: Shocken Books,
1984), p. 157. Here are more likely reasons why the public is cooling off about unlimited
and uncritical funding of science projects:

The Reagan Years: Defense vs. Nondefense R&D (in billions)

	FY 1980 Actual	FY 1988 Estimated	Percent Change	
			Current $	Constant $
Defense R&D	$15.0	$40.3	169	83
Basic Research	0.6	0.9	64	11
Applied Research	1.9	2.6	38	−7
Development	12.5	36.7	194	99
Nondefense R&D	$16.7	$18.8	13	−24
Basic Research	4.2	8.6	107	40
Applied Research	5.0	6.5	29	−13
Development	7.5	3.7	−50	−66

The Reagan Years: Spending for Basic Research by Federal Agency (in millions)

Agency	FY 1980 Actual	FY 1988 Estimated	Percent Change	
			Current $	Constant $
NIH	$1,644	$3,855	134	59
NSF	830	1,438	73	17
Defense	552	901	63	11
Energy	523	1,172	124	52
NASA	559	1,074	92	30
Agriculture	280	471	68	14
Other	328	613	87	26
Total	$4,716	$9,526	102	37

Source: AAAS and "OMB Data for Special Analysis J, FY 1989 Budget."

While I suspect that the British decline of support for academic science relates to the severe pruning of the whole post-sixties university growth phenomenon and a return to the pre-sixties Oxbridge near-monopoly on elite higher education than it does to diminished faith in science, what I am using in this incident is its index regarding a certain lack of maturity within large segments of the science community regarding criticism. While institutional science can rightfully pride itself in its *internal* standards of criticism (directed at the critique of experiment, of data collection, of research design, calculative and deductive reasoning, etc.), it is not always so mature with regard to self-reflexive criticism.

In 1977 Langdon Winner published his book *Autonomous Technology: Technics-out-of-Control as a Theme in Political Thought*; although the title itself indicated that he was addressing a way of interpreting technology, uncritical readers and reviewers often took the result as anti-technological. By the time he wrote *The Whale and the Reactor*, which appeared in 1986, he had learned the following lesson:

Leakage from Britain's plant at Windscale has been described as unstoppable, and the Irish Sea is now claimed by some to be the most radioactive in the world. Some waste products will remain lethal for tens of thousands of years and the technology for their disposal has yet to be developed.

If it were literary criticism, everyone would immediately understand that
the underlying purpose is positive. A critic of literature examines a work,
analyzing its features, evaluating its qualities, seeking a deeper
appreciation that might be useful to other readers of the same text. In a
similar way, critics of music, theater, and the arts have a valuable, well-
established role, serving as a helpful bridge between artists and audiences.
Criticism of technology, however, is not yet afforded the same glad
welcome. Writers who venture beyond the most pedestrian, dreary
conceptions of tools and uses to investigate ways in which technical forms
are implicated in the basic patterns and problems of our culture are often
greeted with the charge that they are merely "antitechnology" or
"blaming technology." All who have recently stepped forward as critics in
this realm have been tarred with the same idiot brush, an expression of
the desire to stop a much needed dialogue rather than enlarge it.[12]

In the "*Nature* controversy" one can see that the same unself-reflexive
tendencies occur within some science circles. One must be careful
here to qualify any too-wide a claim, since even in this controversy,
scientists as well as philosophers reacted critically to the Theocharis-
Psimopoulos article.

Yet unlike in the arts and letters, the role of criticism—particularly
beyond the narrower and internal limits of self-regulation—is not yet
fully assimilated to science and technology as institution. The history
and philosophy of science *should* be regarded as analogous to art and
literary criticism, and the historian and philosopher as professionals
who provide insight and often deeper appreciation, as well as corrective and negative criticism. That clearly was the aim of the four leading
New philosophers of science portrayed in *Nature* in a photographic
display titled, "Betrayers of the truth? Left to right: Karl Popper, Imre
Lakotos, Thomas Kuhn and Paul Feyerabend."

Before leaving this close-up controversy, one more suggestive
point is revealed in the incident. The very call for (undefined) defense
of "truth" and "objectivity, " in the context of preserving a mythos
that would effectively uncritically support science research, has come
under scrutiny in the sociology of science. The claim for exemption
from criticism can itself be ideological:

We might not be too surprised if the members of society keenest to
subscribe to the notion of scientific neutrality were the scientists
themselves. Many practising scientists, unaware or dismissive as they
are of philosophy of science, cling to an idealized conception of their
profession and propagate a view of "scientific truth" which implies
complete certainty, objectivity and detachment. Such a view may be
held in the full knowledge that many kinds of science can be practiced
only by virtue of financial support provided by governments or
industrial companies with goals which are frequently unclear, and

12. Langdon Winner, *The Whale and the Reactor*, p. xi.

almost always directed by political or economic interests. Incompatible as these two positions may seem, it yet remains true that the prestige and authority of science is such that they are widely accepted, more or less unthinkingly, by the public at large. To some degree at least, science has usurped the territory formerly held by religion.[13]

If the science/philosophy-of-science relationship and traditions of debate concerning the interpretation of science and technology are fairly recent, a much older debate stretching back to the very beginnings of modern science, the religion/science relationship and debate, must at least be mentioned. And although I shall not extensively address this issue, one can note that often contrary mythoi are at contest here, too. Textbook-level expressions of at least one dominant variant of science mythos relates to its own version of demythologization. Indeed, the already-examined theory of progressive revolutions (Copernican, Darwinian, etc.) is in itself an instance of this mythos. The science myth seeks to displace from the religious myth what science terms superstition, which is to be replaced by the rational and usually reductive metaphors of a mechanical or electronic type. But in so doing, science also displaces the possibility of a certain type of valuation preserved in the religious myth. Thus there continues a contestation between these two myth types.

Much of interest and importance has issued from this debate, particularly regarding key myth-metaphors concerning the universe, the idea of the human, etc. While the outcome ranges from the current extremes of a fundamentalist reaction to science itself—symptomatized in the current "creation science" debates—to accommodations between science and religion in more liberal contexts, the overall thrust has clearly been one of increasing secularization in those countries where science-technology as institution takes deepest root. That is part of the institutional non-neutrality of technological culture.

A third and, to my mind, more interesting contemporary critique arises out of philosophical and historical *revisionism* concerning how science operates and is to be understood. An interesting side effect of this debate contrasts the understanding of the process of history by some philosophers and historians with that of some scientists who sometimes wish both to control that history and to keep it within the matrix of "great men, great events" narratives.

While not all scientists interested in science history can be accused of historical naiveté, the dominant form of science-originated histories (as evidenced by the book reviews in *Science*) is that of biographical or autobiographical-like narratives. In itself, there may seem no harm to such an approach. But insofar as it serves to perpetuate one of the standard myths about science, such accounts need, at the least, more critical supplement; at the most, serious demythologizing.

---

13. Steward Richards, p. 148.

That myth is the hero-scientist, working alone or at least individual-istically making some great discovery, cast in some dramatic and roman-tic setting. Contemporary Big Science does not work this way, according to most social historians, either now or in the past, except in a few cases. But the perpetuation of this myth structure does play a role:

> Thomas S. Kuhn, for one, pointed out that young persons must be recruited into science through implicit promises of heroic adventures on the frontiers of knowledge; they would not be enticed by learning that 99 percent of them will spend their lives merely solving the "normal science" puzzles that constitute the vast bulk of research today. Nor, we can add, would they be enticed by the prospect of a "good job on the assembly line" in the production of scientific knowledge which is the social form within which normal science is practiced.[14]

Yet, even this "standard view mythology" can allow peeps through to something rather far from than the intentionality of that view. Still us-ing that narrative shape, James Watson's *The Double Helix* revealed more than most persons in the sciences wanted—as reviews indicated. In his "great men, great events" approach, Watson's telling of the Watson, Crick, Wilkins race for a Nobel prize for a DNA code cracking revealed a warty, human underside to the process. Not only were the principals often competitive, jealous, often verging on the borders of cheating and reluctant to give credit, but *The Double Helix* showed the overt and direct form of male chauvinism that has become the tar-get of the most pointed of contemporary science critics, the feminist critics.

Watson's condescending attitude towards "Rosie" shows through. It was from Rosalind Franklin's crystallographic imagery (again, image-technology) that Watson got his insight—but he repeatedly also reveals his chauvinism and condescension to our women in action. Ruth Bleier has strongly pointed up this trait:

> There could be no finer demonstration of the centrality of gendered metaphors in science [which continue as] . . . misogynist arrogance [which] has continued to thrive among Nobel laureates in science. James Watson was the wonderboy from Harvard when . . . he won the Nobel Prize in 1962 for describing the double helical structure of DNA . . . after illicitly and secretly viewing Rosalind Franklin's unpublished crystallographic images of DNA.[15]

That same "wonderboy," as late as 1985, complaining of the move to regulate genetic engineering, was quoted as saying:

---

14. Sandra Harding, *The Science Question in Feminism*, p. 69.
15. Ruth Bleier (ed.), *Feminist Approaches to Science* (Madison: University of Wiscon-sin Press, 1986), pp. 6–7.

One might have hoped that the Republicans would have been more sensible about regulations, but they were just as silly as the others. . . . The reason is that the White House receives its advice from people who know something about physics or chemistry. The person in charge of biology is either a woman or unimportant. They had to put a woman someplace. . . . [16]

Were one to shift to the social historians, a different initial picture would emerge. Derek de Solla Price, our previous interpreter of Galileo, was also part of the revisionist revolution of the early sixties with his now well-known *Little Science, Big Science* (1963). There he began to show some of the social and historical features of what is now commonly known as Big Science. He showed that beginning in the nineteenth century, the institution of science began to grow exponentially, and science literature with it. But it also began to be truly corporate science, indexically illustrated by the way in which authorship of publications changed. Today it is *rare* for any science article to have a single author (apart from opinion or state-of-the-art summaries by leaders in the field—even our aforementioned *Nature* commentary was co-authored!). And in spite of the equally important outline of what Price called Invisible Colleges of the small group of elite leaders and publishers in any given field, co-authored publications often were the result of an army of laboratory workers with only scant and distant control of the "P.I." (principal investigator). This corporate structure of publication and research is very far from the Great Man histories still favored by many individual science writers. Evidence of the too-frequent distance of the elite P.I. from the laboratory and data-gathering work of the "factory" is the growing phenomenon of data fudging and other forms of corporate science cheating, which now play a constant role in the discussion of many leading science magazines.

Watson's *Double Helix*, however, remained within the Great Man tradition, although it was autobiographical enough to reveal warts as well as presumed intellectual heroism.

Reaction from within the science community to this dirty-laundry approach was often strongly negative. It cast too much human foible and weakness upon the process which, mythically, many felt needed to be more romantically portrayed. This naive historical perspective did more to humanize the science operation than many other works still being published; but it also telegraphed a sense of raised contingency appropriate to perceiving science as but one more human endeavor within which culpable humans work.

Nor should one neglect the most revolutionary of the revisionists, Thomas S. Kuhn himself. Although Kuhn probably did more to demythologize the *philosophers'* interpretation of science than he did the

---

16. Ibid., p. 7.

history of science alone, he clearly reformulated the way discoveries
frequently occurred. Even if his own version of that history remains
partially steeped in a somewhat idealized version of theory primacy,
he was able to shift attention to important elements of perceptual
praxis within science. This shift in science history is by now well
known, and I shall not follow it further here. But it did play part of the
role of demythologization, which I am supporting as a necessary one.

I, too, have followed two other strands of this revisionist history in
this book. One relates to the revisionism of the standard view that
modern technology is "applied" science or derivative from science.
The more radical revisionism of that view was Heideggerian; it inverts
the usual view and sees science as the necessary result of a technological wing of science. In a more moderate fashion, a similar view arises
from White's version of the Medieval technological revolution which
sets the stage—including the power metaphor—for modern science.
At a still more moderate level, but an accumulative one, Rachel
Lauden makes a similar point:

> Recent attacks on the concept of technology as applied science have
> employed two strategies, one empirical and one analytic. On the
> empirical front, historian after historian has chronicled episodes in the
> development of technology where the major advances owed little or
> nothing to science. Whether one takes steam power, water power,
> machine tools, clock making, or metallurgy, the conclusion is the same.
> The technology developed without the assistance of scientific theory, a
> position summed up by the slogan, "*science owes more to the steam
> engine than the steam engine owes to science.*"[17]

That revisionism was contextually appropriate for my focus upon
technology. But indirectly, it is part of the demythologization of science as the most important of human cultural knowledge-gathering
activities—the claims here have simply been too strong and need to
be moderated to return science to its human scale.

The second revisionist strand, again not unique to this perspective
but arising more directly from the view of science as both embodied
in technology (instrumentation) and as contextually located, is and has
always been deeply embedded in its concrete connections to technologies in their wider roles. These include, minimally, tool making and
instrument development (de Solla Price) and, more maximally, the
engineering needs of the military and industry. Except for the myth-
makers of the nineteenth century and the positivist theoretical idealiz-
ers of the early twentieth century, I see no need to have ever
conceived of science in any other way. At the very least, this non-
purist view of science as a social institution helps to keep it within balance as *a*, not *the*, privileged human knowledge activity.

---

17. Rachel Lauden, *The Nature of Technological Knowledge*, p. 10.

The revisionism that would see science from the beginning more closely wedded to technology may be debated with regard to historical moments; but in the twentieth century, there is little doubt that some form of science-technology institution is the dominant mode. Even a most conservative view would hold that

> The convergence of science with technology on a large scale began, then, with the appearance of the industrial research-and-development laboratory towards the end of the nineteenth century, a trend started in Germany and followed rapidly in the United States and then in Britain. . . . Industrial symbiosis thereafter became self-perpetuating because it was increasingly hazardous for governments and commercial companies to be left behind in the competitive race for technically new processes and products. . . . The results in the twentieth century of this near-union of science and technology are too numerous and too well known to need repeating.[18]

These interwoven forms of revisionism, of course, reverberate positively with the phenomenological and hermeneutic critiques originating in Euro-American philosophical traditions. Those who place science as an institution within a field and see it as a form, even if a unique form, of human praxis are those who have guided the inquiry to this point.

One of the newest entries into the field of demythologizing critiques, however, has been the appearance of feminist criticism. This strand of critique is a particularly interesting one. Not only is it distinctively postmodern but it has isolated a theme revolving around gender perspectives, which have given it a perspective very focused and yet reaching across cultures and histories. To demythologization should be added the project of "demasculinization" or, positively put, *gender pluralization*.

Special attention should be drawn to this mode of critique because feminism, in one respect, cuts across all previously noted cultural combinations. Whatever the variety of cultural stratgies for genders—and if feminists are right, by far the dominant strategy has been one favoring males—then this new perspective is to be engendered by investigating sex-role perspectives at both micro- and macro-levels. Within science, on a small scale and in specific targeted areas, feminist critiques and perspectives have already succeeded in revolutionizing subfields. One of the most suggestive has been the area of primatology in its behavioral dimensions. However Watson sees it, there is a larger percentage of women in the biological sciences than in the physical and chemical sciences (which, he correctly points out, have had the inside track in political arenas), and within some of the biological sciences—again pointing to primatology—women scientists have come to the fore. Jane Goodall and Dian Fossey have captured

---

18. Steward Richards, *Philosophy and Sociology of Science*, p. 118.

the popular imagination with their respective studies of chimpanzees and gorillas.

What may not yet be fully appreciated is the way feminist perspective has transformed primatology with respect to sexual roles among primates. One example comes from Sarah Blaffer Hrdy's work on lemurs. Attacking the long Darwinian tradition, which interpreted the female of the species in sexual selection as passive, shy, and coy (obviously appropriately linked to Victorian values for the female), Hrdy established what could be called a counter-version of sexual selectivity among lemurs mimicking more closely the liberated, late-modern female of the present.

Hrdy argues, "the initial dichotomy between actively courting, promiscuous males and passively choosing, nonandrous females dates back to Victorian times. 'The males are almost always the wooers,' Darwin wrote."[19] In her own studies and observations, Hrdy says, "I would argue that a polyandrous component is at the core of the breeding systems of most troop-dwelling primates: Females mate with many males, each of whom may contribute a little bit toward the survival of offspring."[20]

Hrdy's studies are not only feminist, they are part of the previously mentioned diminishing of distance between us and our animal neighbors, although here it revolves around gendered issues. She along with others, by taking a different perspective, has clearly demasculinized the field of primatology. The coy female has definitely disappeared from species as diverse as Drosophila and the felines, from shiner perch to savanna baboons, the females of which all seem to engage in active courting and less-than-shy behavior.

A cautionary qualification must be entered regarding the histories of such "paradigm" changes. Donna Haraway, a peer biologist turned historian of science, critically recognizes the need for a careful hermeneutics of all such shifts. Recognizing that the feminists who have made a major impact, and who in sexual selection areas of primatology research are now probably the dominant researchers, they also must face the ambiguities of reading the human via the animal (text). Haraway shows the historian's more sophisticated critical attitude towards science, which I have claimed is often more lacking from within the science community itself.

Haraway underlines the literary, symbolic character of science itself in its latest attempts to get at the tenuous line between the animal and the human in primates and of the new role being established by women in primatology. In her "Primatology is Politics by Other Means," Haraway notes the advances made:

19. Sarah Blaffer Hrdy, "Empathy, Polyandry, and the Myth of the Coy Female," *Feminist Approaches to Science*, p. 118.
20. Ibid., p. 125.

> Until very recently in the history of primatology, virtually no women had the status of scientist in this field. . . . [But now] these woman have made a major difference in scientific constructions in primatology of what it means to be a female animal, and so of what it means to be a man or woman in societies for which the social construction of animal is part of the social construction of human.[21]

Yet it is precisely because of this borderline status that primates are hard to "read." Haraway points out that "Westerners have access to monkeys and apes only under specific symbolic and social circumstances. . . . It is very hard to stabilize the truth about monkeys and apes."[22]

As a post-Kuhnian historian of science, Haraway rightly notes that even what is called a "fact" is context- or field-dependent; but more, it is field-dependent in the sense of being "story-laden":

> *Values* seems an anemic word to convey the multiple strands of meaning woven into the bodies of monkeys and apes. So, I prefer to say that the life and social sciences in general and primatology in particular are story-laden; these sciences are composed through complex, historically specific story-telling practices. Facts are theory-laden; theories are value-laden; values are story-laden.[23]

This is now standard "new" or post-Kuhnian philosophy of science. (It is also good phenomenology.)

With the now sharpened hermeneutic caution, we can return to Hrdy's example. We can see that without in any way demeaning her field observations about lemur behavior, her "facts," Hrdy has done more; she has told a new story, provided a new context of interpretation. Again, Haraway:

> Rarely will feminist contests for scientific meaning work by replacing one paradigm with another, by posing and successfully establishing fully alternative accounts and theories. Rather, as a form of narrative practice or story-telling, feminist practice in primatology has worked more by altering a "field" of stories or possible explanatory accounts. . . . Every story in a "field" alters the status of others. The total interrelated array of stories is what I call a *narrative field.*[24]

That is what Hrdy had done with lemurs, but in this case it is also what may be called a positive "demasculinization" of science. I do not mean to cast a negative or pejorative tone upon the masculine except

---

21. Donna Haraway, "Primatology is Politics by Other Means," *Feminist Approaches to Science*, p. 78.
22. Ibid., p. 79.
23. Ibid., p. 79.
24. Ibid., p. 81.

in its tending to chauvinism (as in the Watson example above), but to indicate that the alteration of the narrative field allows the multiplicity of gender aspects to be reassessed and re-entered in a more complete picture. The demasculinization, ideally, is the first step towards *gender pluralization*, a more appropriate postmodern condition.

Yet the previous history of science can be seen as a progression of masculinization. Sandra Harding, in one of the most comprehensive and telling of the feminist critiques of science, makes the following claim:

> When Copernican theory replaced the earth-centered universe with a sun-centered universe, it also replaced a woman-centered universe with a man-centered one. For Renaissance and earlier thought within an organic conception of nature, the sun was associated with manliness and the earth with two opposing aspects of womanliness. Nature, and especially the earth, was identified on the one hand with a nurturing mother—"a kindly, beneficent female who provided for the needs of mankind in an ordered, planned universe"—and on the other with the "wild and uncontrollable [female] nature that could render violence, storms, droughts, and general chaos." In the new Copernican theory the womanly earth, which had been God's special creation for man's nurturance, became just one tiny, externally moved planet circling in an insignificant orbit around the masculine sun.[25]

I do not know if, as some feminists claim, the values of care, nurture, empathy are either accidentally or intrinsically more feminine than the values of protection, domination, or control, which are sometimes alleged as male values; but it is clear that a science dominated by the latter clearly needs a heavy dose of the former if balance is to be attained and even more if the earth is to be conserved. For that reason alone, one of the interstitial tasks for the present must be the further demasculinization of science within and alongside the task of demythologizing its false mystique.

Did Galileo know, when he confirmed the Copernican theory through his new artificial revelation, that he was also decentering the earth as our source of nurture? But then, Galileo must serve here as our "primate" in that it is through Galileo—or the multiple Galileos we have already seen—that the story of science is told and retold. Demythologization and associated demasculinization are themselves part of the new narrative that is needed to alter the current field of stories.

If science is indeed the conceptual tool of technological civilization, then its remaking into a tool enhancing the care and nurture of earth as our inherited habitat belongs both to the re-situation of science within society and to the task of developing a worldwide conservation ethic. Both are interrelated.

---

25. Sandra Harding, *The Science Question in Feminism*, p. 114.

C. GALILEO IN THE KITCHEN

If the need is for a worldwide sensibility for conserving the earth, and one means of working towards that end entails demythologizing and de-masculinizing extant science, what of technology? From what has preceded, it will be clear that I have not argued for any of the sometimes popular programs related to alternative technologies, either "appropriate," "small is beautiful," or even for decentralized over centralized systems or subsystems. While I am certainly not opposed to any of these directions that would correct the abuses and the large negative oscillations found in many current trajectories, I am skeptical about such panaceas both because of the misconceptions revolving around "control" issues and because none seems plausible from within current vectors.

Those vectors include the recently passed defense budget for the United States at a $300 billion figure (1989). This magnitude clearly indicates *which* technologies receive largest favor. To resuscitate the framework from the field of stories about technologies calls for a much broader and deeper change than is possible here, yet that is what philosophers, historians, and humanists must continue to strive for. Such a change must also occur within the dominant community of technological science itself.

If Galileo has been our "primate" through whom we "read" contemporary science, can there be a second Galileo through whom to "read" a postmodern science?

Galileo II finds himself in the kitchen after his day in the physics lab. As a liberated husband, he shares in preparing dinner for Eve, his wife, and Adam, his young son, who will arrive shortly from work and school. He likes his culinary task not only because he enjoys the gadgets—food processor, carbon steel knives, copper-bottomed cookware—but because while he works on the varied but healthy menus, he has a chance to daydream. So tonight, as he prepares a first course of *poisson cru*, a French-Tahitian fish dish, to be followed by stir-fried Oriental-style vegetables, he indulges in a bit of wish-fulfillment thinking, which he recognizes skirts on the fantasies of utopia.

Of late, he has been thinking of a midlife career change, typical of postmodern life, since he has wearied of the constant stress of grants applications and the too-often-determined priorities of the grants agencies. He has been a critic of SDI, the space station, and even has doubts about the wisdom of a super-collider, particularly in the light of the multi-billions involved. Moreover, he knows that even amateurs could easily dupe the planned sensor and laser cannon system being dreamed up for the first, unmanned vehicles could do more actual science observation than the second, and likely technological advance could well make the latter obsolete before it is finished.

He wonders what a science world would be if its current hierarchies, so well-entrenched in academia, politics, and the military-industrial complex, were shifted. What if the physical and earth sciences were geared closely to conservation programs—responding to demonstrated public concerns for cleaning up the oceans and the atmosphere? What if the ecological disciplines took precedence over the current rage for biotechnology? And what if the social sciences and even the humanities and arts were funded on anything like the scale of his science?

Carried away with himself, he begins to fantasize a new type of technology assessment that not only include the current minimalistic environmental impact studies but also require an aesthetic and community impact statement. He is well aware that in his university the dominant administration *always* asks if its history and humanities programs have curricular elements relating to the sciences—what if the planned Waste Management Institute were required to have a like connection to the art community and world, the "artistic" dimension of waste disposal? What if all technological development were to depend upon and be funded in relation to its nurturing of environmental, social, and distributive-justice elements?

He had begun to think in this admittedly heretical and clearly wishful way ever since he began to wonder what the world would be like for young Adam. Sometimes he envied the boy, who could with equal enthusiasm enjoy the lizards sunning themselves on the rock wall by the house, the loud and noisy MTV channel, and his toy computer.

Before he can get too carried away—and burn the stir fry!—Eve arrives and reports to him that some of her peers in the Northwest had just finished a study of young men and in the sample poll discovered that the most important single value they rated highest was *fatherhood*. This was the first time in the history of similar polls that fatherhood even made the list! Eve, perhaps to tease him, sometimes argued that in addition to a non-Western culture requirement for the core curriculum there ought to be a required "nurture" set of courses, focusing upon the variant forms of human and animal nurture, to balance the already required "technological literacy."

That argument sent Galileo II off once again as he returned to the kitchen to prepare a dessert of cheese and fruit. What would the funding structure for science and technology be if the ancestors of his namesake—the princes and merchants of that day—were replaced by funding agencies geared to social needs, economically just wealth distribution, issues of human nurture, and conservation of planet Earth? He could not even imagine such a possibility—although to fantasize what kinds of technologies would be invented in such a world might be a nice daydream for tomorrow night.

D. CONCLUDING POSTSCRIPT ON TECHNOLOGICAL SCIENCE

Galileo II is a fantasy. Nor will the changes wishfully dreamed of occur through individual conversions or the nexus of the family. The problems with modern technology are too deep and too entrenched, with centuries-long habits, to change in any easy way. Yet their distortions have begun to appear.

In contrast to the postmodern spirit, technology and technological science remain wedded to large hierarchical structures and, in Big Science, even to what has been analyzed as something akin to a factory system:

> The manager-distributors of science are only a small minority of scientific workers. One source estimates that "some 200-300 key decision-makers—primarily scientists—constitute the inner elite out of a total scientific work force of some two million." Performing almost all of the labor actually required to produce scientific belief are the 1,999,700-or-so technicians in laboratories and workers who manufacture the equipment and materials for scientific inquiry.[26]

What is sad, even tragic, about this state of affairs is that the genius of modern science *is* its technological embodiment. It was through that embodiment, in the picking up and development of instrumentation for experiment, that the entire seeing of the world changed. It revealed to us micro-worlds and macro-worlds not even dreamed of by the imaginers of the pre-modern science histories and cultures. Nor could the eyeball cosmologies begin to elaborate or extrapolate the wonders of the instrumental realism of modern science.

This explosion of knowledge, which is the heritage of modern science, has led to the proliferation of what I have called the "compound eye" of the present. It is the multiplied, refracted vision of an even overheated postmodernism. No more is the *trompe-l'oeil* construction of a mathematized perspective painting from the Renaissance the symbol of the age; rather, it is the repeated compound eyes of the control panels from NASA to CBS to the artistic taking up of these compound eyes within the cinema arts and visual arts. These eyes reveal a multiple, pluralistic world.

The ambiguity of this wonder, though, was the doubled relation to technology. On one side, modern science was from the beginning technologically embodied; but on the other, it was also technologically embedded within the sought-for support systems of the powers of a society become technological. In the era in which the magnification and oscillations of that world have become what they have, this doubled relation becomes the distortion which is now revealed within technological science.

Still, something else is also happening. The very interlinkage of the earth, which high-technology culture has brought about, has become the place for a postmodern sensibility to begin to show itself. No one, I think, would argue that the emergent sensibility has yet any signifi-

---

26. Ibid., p. 72.

cant *power*; but as a possible gestalt shift, it is clearly differently focused than the dominant vision.

Its appearance is sharpest within the artistic, literary, and philosophical communities, whose sensitivities are always aimed at such
shifts, but postmodernistic sensitivity is not limited to this domain. It is
a sensibility that rejects a hierarchical system, which it sees as anachronistic, whether the system be conceptual, as in traditional metaphysics in philosophy (which results in the rejection of foundationalism by
postmodern philosophers), or in the arts (which could be said to take
Andy Warhol's cryptic comment seriously). If everyone is to have his
or her moment of fame—whether for ten or more minutes—this observation is actually a postmodern limit-statement. Each artist, each human activity must take its place beside the others in a non-hierarchical
plurality of democratized moments. That is part of the postmodern
spirit. It is reflected in the shape pluriculturality has taken. And that is
out of one of the primary trajectories of high-technological culture in
the midst of—and in spite of—its anachronistic modern retention of
the hierarchical.

Were a rearrangement of the hierarchies of the sciences possible—and there are some shakings of that hierarchy today through the
challenge of the biological sciences, in contrast to the entrenched
physical and chemical sciences—the result would be merely a new
hierarchy. The trajectory of postmodernism is away from the hierarchical, towards a pluralization. I do not know, of course, whether that
sensibility will become wider or more sedimented than it now is, but it
is operative in the contemporarily apparent way of seeing.

I have argued that this vision is of a compound eye, a vision refracted by a plurality of views, each of which casts perspectival light
upon its objects. That among other things is what the feminist critique
of institutional science has done in high relief. The once accepted traditions of even the history of metaphor have been challenged. And
their appearance today must appear not simply "in bad taste," but
anachronistic. And that, too, is a result of postmodern vision. That the
science establishment is sociologically dominantly male—from mathematics through the physical and chemical sciences and, except for isolated islands, within the biological sciences—stands out in high relief.
Arguments that this situation does not eventuate in a "masculinist" or
gender-viewpoint distortion are increasingly hollow. The "rape of nature" metaphors which have run continuously from Bacon through
contemporary Nobel speeches belie innocence.

In social formation, too, this institutionalization of science remains
parallel to other large corporate entities such as multinational corporations, banks, or—symbolically—the locker room of now Big Football.
Such hierarchical, often authoritarian, and clearly top-down structures
are far from the postmodern.

Were the postmodern spirit to reign among these large entities, a different picture would emerge. The football locker room is, admittedly, not likely to be gender-pluralized in any near future, but a genuine postmodern *Civilization* would not make such an activity, which preserves the aggressive, testosterone-driven male, either the dominant or the sole activity of its repertoire. The point is to enhance the distortion: To have such an activity as a sport, as one activity among others, is one thing; to have made it the model for the investigation and ultimately the "control" of the whole of nature is quite another. Yet the combination of institutional science within its technological culture embeddedness with its multicorporate and multinational sponsors—all of which show the same structural features—is what now relates to the realm of the whale (oceans), the rain forests (the earth), and the skies.

Thus it is of the triumvirate of the symbolic technological tragedies of the late twentieth century; and, in spite of the size of the human group killed, the "Challenger" incident is the most pointed for us and must take symbolic precedence over Bhopal or Chernobyl, whose actual human impact was and continues to be much larger.

The very name "Challenger" reflects Heidegger's much earlier observation: "The revealing that rules in modern technology is a *challenging* which puts to nature the unreasonable demand that it supply [us with itself as a resource-well]."[27] This name, pugilistic, phallocentric, conveys that same sense. And carefully planned through the slick, expensive NASA literature disseminated in virtually every elementary school, the deliberation that took along a teacher-mother to lead a generation of children into the venture was what made "Challenger" the event that it was. And the event echoed at the end of a century precisely what occurred earlier in the century in another hubristic name, "Titanic."

There is also a difference. Whereas the prevailing mythos of the "Titanic" was one steeped in romatic notions of "nature's revenge," the "Challenger" incident reveals more about the designers' and the decision makers' molding of technological science. It was not nature which took the "Challenger"; it was much more the locker–board room overconfidence in the rightness of the task. Nor have any of the deeper lessons been learned. "Discovery," just launched in similar full-media coverage, complete with schoolchildren flown in for the occasion, is a name only slightly nuanced regarding dominance. And the language of "re-entering" the space race once again elicits, for the postmodernly aware, the continual phallocentrism of the process. In fact, owing to previous incident, civilians and women are now even more strongly excluded from the activity. It reverts to a population closer now to that of the locker room than during the period of tokenism previously undertaken.

---

27. Martin Heidegger, *Basic Writing*, p. 296.

To become a postmodern Civilization in which there would be a democracy and pluralization of diverse human activities, such pointedly phallocentric activities would have to be restricted to the limits of games proper. All the more important activities would have to have both gender and cultural pluralities as their distributed basis of action. I am not here suggesting that if we merely correct the gender bias of technological science we will then have solved our problems. That is only part of the problem, although it is an aspect that the cross-cultural perspective of feminist criticism has made appear in sharp relief.

We are back once again to the misplaced question of "control," but it is and remains a question of how one changes a technological *culture*, not simply one of which technologies in the abstract can and should be developed. I recognize, too, that I am perilously close to blaming the embeddedness of Big Science in Big Technology for the distortion while praising the emergence of a pluralistic, postmodern vision of a compound eye—which could be misinterpreted as a "little" technological science. That is not factually the case, since the emergence of a compound eye is itself the result of both a big and, even more, a *networked, multinational technological* science.

The present distortion is an imbalance, possibly even a fatal one, arising out of contrary trajectories existing within the double technological embodiment and embeddedness that has come to be the situation of the contemporary technological lifeworld. Whether, out of the proliferation of perspectives constituting pluriculturality, with its compound eye and new modes of inquiry, there can come sufficient vision and balance to correct the now dominant trends of late modern technology, I do not claim to know.

As of this moment, the largest of all technologically complex systems—the military one—is stalemated. Even while it continues to grow and be developed, its actual use is symbolic and hermeneutic, a dial indicating the comparative weights of the powers who brandish it. Its smaller relatives, of course, find enough employment in local, regional, and terrorist uses. These come close to our very lives with too much frequency. The terrorist bombing of the Pan Am clipper flight 103 out of London on December 21, 1988, occurred exactly one week prior to my own flight home on its sister ship on the 28th.

In a related way, although I have not addressed it systematically here, the same modern anachronism has begun to appear in corporate life. The United States, concerned that it is falling behind both competitively and in technology development, has begun to realize that the model of a hierarchical corporation, particularly one not engaging its workers more relationally and democratically, has become something of another dinosaur form. The newer and more dynamic forms of industry now taking shape—particularly Japanese—are even more high-tech than their predecessors, but they are also embedded in a different social-cultural form. They are driven by a relational, consensus operation. They are almost "socialistic" in terms of worker-support plans. They no longer follow the single-job, assembly-line pattern of a now-outdated industry. (This is not to claim that they are also gender-pluralized, concerned about the environment, or conservationist in direction!)

The technology of the compound eye also grows and is being developed as the various forms of networking reach into ever-new crannies of the globe, bringing with them the awakening to non-neutral pluriculturality. Whether this growth will stimulate a new tolerance and postmodern taste for a proliferation of all alternatives also remains to be seen. Yet if the multiple vision of the compound eye is a transitional indicator of a movement beyond the modern—even if its ultimate shape is yet unclear—its form of vision can be partially described. That vision is multiple, aware of multistability, refracted, and perspectival. At present, its field of vision, not yet fully gestalted, still appears something like a bricolage field. It is strewn with cultures and culture-fragments. And the popular culture visionaries have entered the field with a sense of playfulness, for they see in the field a condition for *invention*.

Can there be an "invented" culture? If so, it will not be the result of an individual but of an ensemble. The playfulness found in and among the artists of the image-technologies who have discovered this bricolage is often suggestive. For example, the movie *Koyaaniskatsi*, utilizing in typical postmodern fashion such techniques as time-lapse photography, also draws from Hopi time senses (a cultural fragment) . . . to show us something about our current situation within the technologically textured environment. Here is a technologically derived art form, playing back reflexively upon its sources to give us a glimpse of our own lifeworld.

Even within the rapid survey of human-technology relations and of culture fragments revealed here, one could project other such culture inventions. For my part, it would not be hard to imagine a possible world in which the superior environmental sensitivities of inland Australian Aboriginals were bound to certain kinds of even high-technology developmental trajectories. I showed that the existential structures of human-technology relations were the same for scientific and for *musical* instruments—although the respective "worlds" projected by each were radically different. We have explored and are exploring the profusion of instruments relating to the first context; we are very conservative with respect to the latter. In a high-technology, "Aboriginally" constituted world, the latter might be more interesting and fascinating as a trajectory than the former.

The bricolage culture invention, however, is at most a fantasy *variation*; yet such perspectival variations are also part of the multistability of postmodern vision. They bespeak a certain lightness to vision not often understood or appreciated as part of the postmodern. It is a "Nietzschean" lightness. What is taken as a relativism or nihilism—one reading of the proliferation of choice characterizing the heightened contingency of the postmodern technological world—is negative only if an ultimately weighty choice is considered the only valuable choice. Yet experimentation, invention, is not always or even often so weighty.

The Heideggers and their followers who claim that only a god can save us are perhaps forgetting that their predecessor, Nietzsche, preferred—if there were to be any gods at all—gods who danced. Perhaps the god who saves us must be the god who dances? Yet dance, too, is an attainment. A dancer's lightness is attained, not the result of a spontaneous orgy of enthusiasm. Both the gravity-defying leaps of Western ballet and the rich, stylized gestures of traditional Indian dance can only occur through the long, technical apprenticeships that make dance appear to have lightness.

The now high-technology texture of the lifeworld is one in which the proliferation of the possible is diverse, multistable, and often both confusing and dangerous. It remains the task of the inhabitants to cultivate the right weight and lightness of movement to maintain a balance within that world. We have not yet done that, but it still may be possible to learn the movements.

Index